CHEMICAL VAPOR DEPOSITION OF TUNGSTEN AND TUNGSTEN SILICIDES

CHEMICAL VAPOR DEPOSITION OF TUNGSTEN AND TUNGSTEN SILICIDES

For VLSI/ULSI Applications

by

John E.J. Schmitz

Thin Film Division
Genus, Inc.
Mountain View, California

Library of Congress Catalog Card Number: 91-18643
ISBN: 0-8155-1288-0
Printed in the United States

Published in the United States of America by
Noyes Publications
Mill Road, Park Ridge, New Jersey 07656

10 9 8 7 6 5 4 3 2 1

Library of Congress Cataloging-in-Publication Data
Schmitz, John E.J.
        Chemical vapor deposition of tungsten and tungsten silicides for
    VLSI/ULSI applications / by John E.J. Schmitz
             p.   cm.
        Includes bibliographical references and index.
        ISBN 0-8155-1288-0
        1. Integrated circuits--Very large scale integration--Materials.
    2. Tungsten. 3. Vapor-plating. I. Title.
    TK7871.15.T85S36   1991
    621.39'5--dc20   91-18643
                                            CIP

To Pieternel and Lucas

MATERIALS SCIENCE AND PROCESS TECHNOLOGY SERIES

Editors

Rointan F. Bunshah, University of California, Los Angeles *(Series Editor)*
Gary E. McGuire, Microelectronics Center of North Carolina *(Series Editor)*
Stephen M. Rossnagel, IBM Thomas J. Watson Research Center *(Consulting Editor)*

Electronic Materials and Process Technology

(continued)

"

Ceramic and Other Materials—Processing and Technology

Related Titles

PREFACE

The acceptance of the chemical vapor deposition of tungsten (CVD-W) is such that it is finding its way more and more into high volume production of ULSI circuits. Unfortunately, bringing up a CVD-W process is not a trivial exercise. New equipment configurations (cold wall reactor), new deposition chemistries (not to mention the chemistries of precleaning and/or etching), adhesion layers, etc. all contribute to the complexity of this process. In addition, the maintenance of a tungsten process in terms of reactor cleaning, maintenance and trouble shooting requires a solid background in CVD-W technology.

Extensive literature has been published on blanket and selective CVD-W, in which a vast amount of (sometimes conflicting) information can be found. What is clearly needed is a book where all relevant and pertinent material is gathered in a condensed format. It is the intention of this book to provide such a compilation of the literature with emphasis on the material which has appeared in the last 10 years. In addition, unpublished material obtained in the laboratory of the author is included. After reading this work, the reader will have all the necessary background to bring up, fine tune and maintain successfully a CVD-W process in a production line. Others seeking a quick overview of the current status of CVD-W will also benefit from this book.

The nine chapters of this book can be read in any order. No background other than basic physics and chemistry is assumed. Where appropriate, rule of thumb calculations are included to increase further insight into the subject. The author has also provided personal opinion and insight on certain subjects where appropriate.

Chapter I gives a description of the driving forces behind the introduction of CVD-W in IC manufacturing. Chapter II treats the issues of blanket CVD-W for plug applications. In this chapter the etch back of blanket tungsten is also briefly summarized. Selective tungsten is described in chapter III with the emphasis again on plug applications. Chapter IV compares the benefits of selective and blanket CVD-W for plug applications. Another important application of blanket CVD-W, namely that of the use of tungsten as interconnect material, is extensively discussed in chapter V. Important properties of the gases and chemistries used and of tungsten itself are evaluated in chapter VI. Chapter VII is especially important because it treats the principles of cold wall reactors where much attention is paid to wafer temperature and its effect on the process. Chapter VIII lists several subjects which might become important future applications of CVD-W but are now still in the R&D stage. Additionally, some alternative plug processes are discussed.

The chemical vapor deposition of tungsten silicide (WSi_x) is also covered in this book in chapter IX. This material was included since the chemistry and equipment are so similar to blanket tungsten. Additionally, it allowed the coverage of the SiH_2Cl_2 based tungsten silicide process which is relatively new today.

For the convenience of the reader a comprehensive reference list of over 260 references is included at the end of the book. The literature references are grouped according to their subject. In addition, a subject and an author index will be found which facilitates the use of the book as a reference tool for CVD-W and CVD-WSi_x.

A statement about the units used in this book is in order. The unit system as used in each specific piece of literature under discussion is maintained.

Sunnyvale   J.E.J. Schmitz
March, 1991

ACKNOWLEDGMENTS

During the preparation of this book many people were consulted for advice or asked for original SEM micrographs. The following persons have to be mentioned: Larry Bartholomew, Ray Chow, Russell Ellwanger, Janet Flanner, Clark Fuhs, Dr. Mart Graef, Dr. Albert Hasper, Sien Kang, Dr. H. Korner, Gareth Patten, Dr. Ivo Raaijmakers, Dr. Ed Rode, Steve Selbrede and Dr. Evert van de Ven. Special thanks to Jim Dodsworth and Norm Zetterquist who carefully read the manuscript and gave many suggestions and to Doree Swanson who helped with the preparation of the manuscript. Thanks also to the Genus executive management for the support given during the period of preparation of the manuscript, especially William W.R. Elder.

The Electrochemical Society, The Materials Research Society, Wiley and Sons Inc., Lake Publishing Corporation, The Institute of Electrical and Electronics Engineers, Inc. (IEEE), Solid State Technology, Elseviers Science Publishers BV, Cahners Publishing Co. and The American Institute of Physics graciously allowed the reprint of numerous pictures from their publications.

Finally I would like to thank Noyes Publications, in particular George Narita, for the support given and for his consideration of the viability of this publication.

NOTICE

To the best of the Publisher's knowledge the information contained in this book is accurate; however, the Publisher assumes no responsibility nor liability for errors or any consequences arising from the use of the information contained herein. Final determination of the suitability of any information, procedure, or product for use contemplated by any user, and the manner of that use, is the sole responsibility of the user. The book is intended for informational purposes only. Tungsten deposition raw materials and processes could be potentially hazardous and due caution should always be exercised in the handling of materials and equipment. Expert advice should be obtained at all times when implementation is being considered.

CONTENTS

CHAPTER I

INTRODUCTION

1.1 SCALING DOWN

There are two basic reasons for the ongoing increase of component integration in integrated circuits (IC's):

(a) better performance: the smaller size of the devices (ie. transistors, diodes etc.) often results in a better performance and higher speed of these components and

(b) cost savings: more components can be integrated per unit area or, from another perspective, the size of the IC's can be considerably smaller, thus allowing more of them on each wafer. It is possible that the latter provides higher yields because the risk of particle contamination resulting in defects is lower. Higher yields obviously equate to lower cost per die.

When design rules invade the sub-micron regime, new process problems will occur. These problems are partly caused by the increased aspect ratios which are inherent to sub-micron design rules. For example, low temperature oxides such as SiH_4/O_2-LTO or plasma enhanced Si_3N_4

tend to result in void formation in the dielectric layer when the aspect ratio becomes larger than 1. Another example of a common problem associated with the deposition of thin films on sub-micron features is that the step coverage of sputtered aluminum is not acceptable in sub- micron contacts (see below).

The general approach to reduce problems associated with high aspect ratios is planarization of the steps. Planarization of contacts (or vias) can be accomplished by filling them with a conducting material. The main emphasis of this book is to show how this can be achieved by Chemical Vapor Deposition of Tungsten (CVD-W) either in the blanket or in the selective mode. In addition, other important applications of CVD-W will be

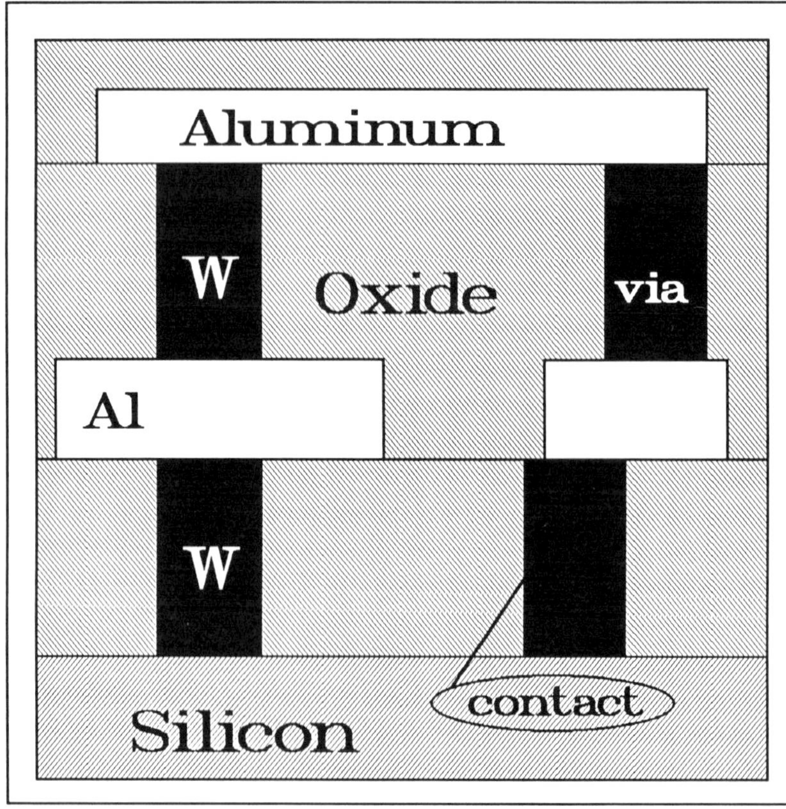

Figure 1.1. Cross section of a multi-level metallization system. In this situation the contacts and the via's are already filled.

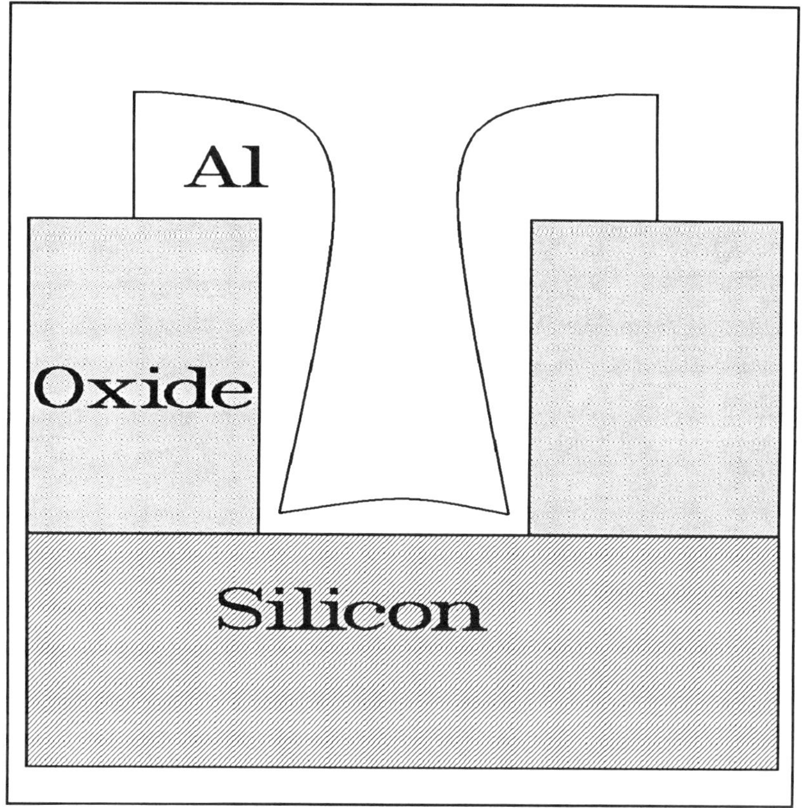

Figure 1.2. A high aspect ratio sub-micron contact filled with sputtered aluminum. Note the thinning of the aluminum at the side walls.

discussed as well as film properties and cold wall reactor fundamentals.

1.2 ELECTRICAL CONTACTS

An IC design with multi-level metallization contains at least two types of contacts:

a) the contact to the active areas hereafter referred to as "contact" and

b) the connection between two overlying metal layers hereafter named as "via" (see figure 1.1).

In most types of IC's, the contacts can end on n^+ or p^+ mono-crystalline silicon, poly-silicon, various types of silicides, and other materials such as TiN. One of the most important properties of the contact and the via is the contact resistance (R_c):

$$R_c = \partial V / \partial I \qquad (1.1)$$

The determination of R_c is not trivial and care should be taken that the appropriate device (Kelvin) is used (see for more details chapter III). Values of R_c found in the literature for contacts direct to silicon are in the range of 10^{-7} Ωcm^2 and vias are in the range of 10^{-8} Ωcm^2.

1.3 DEVICE RELIABILITY

Consider the ramifications when a sub-micron, high-aspect ratio contact will be filled in the conventional way using sputtered aluminum (see figure 1.2). When the step coverage is only minimally acceptable, the aluminum can still provide continuous conductance and electrical contact. In fact, R_c from such a contact, as measured from a Kelvin structure, can still be excellent under such conditions. Two problems, however, remain with this approach:

(a) During current passage a very high current density is seen at the bottom of the contact hole where the sputtered aluminum tends to be thinnest. This can result in Joule heating and, even more serious, in increased electromigration of the aluminum in the contact. Eventually this electromigration can lead to an open contact and the loss of the integrity of the circuit or a dead circuit.

(b) Due to the poor step coverage of the aluminum layer, very high aspect ratios usually remain after aluminum deposition. Since in the following step, a dielectric layer must be deposited, void formation in this

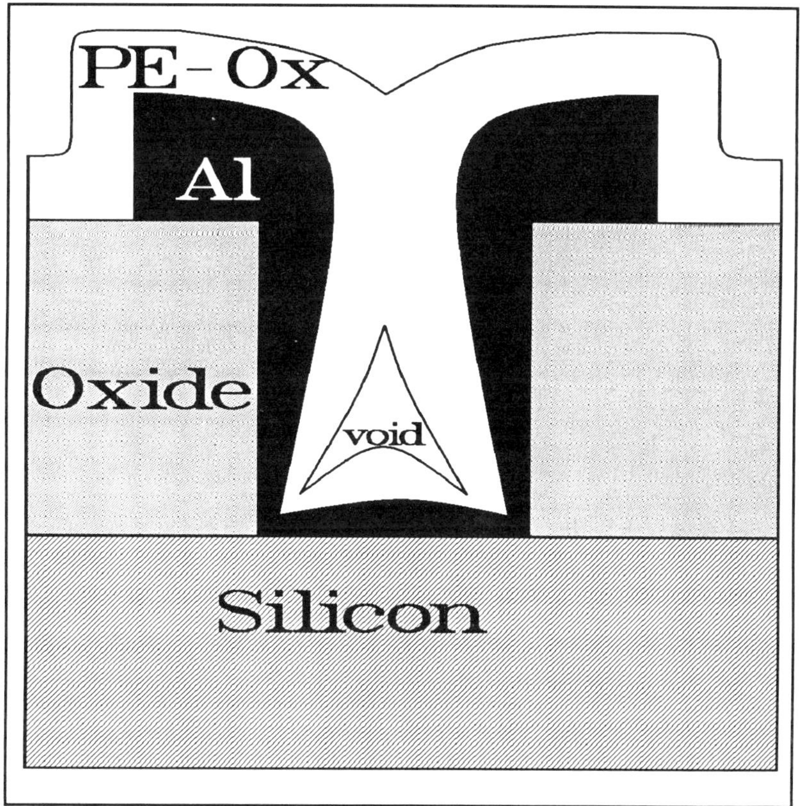

Figure 1.3. The same situation as in figure 1.2 but now after the deposition of a dielectric layer (for example plasma enhanced CVD-oxide).

layer can easily occur (see figure 1.3). Such voids are generally considered to be a reliability hazard. For instance, crack formation or enhanced aluminum mobility can occur. Moreover, when a resist etch back (REB) is used for dielectric planarization such voids can cause intra-metal shorts.

Although several attempts have been made to improve the step coverage of sputtered aluminum, the results have not been optimal because other properties (such as electromigration resistance) of the aluminum were degraded. Clearly in ULSI there is a need for a contact/via planarization method.

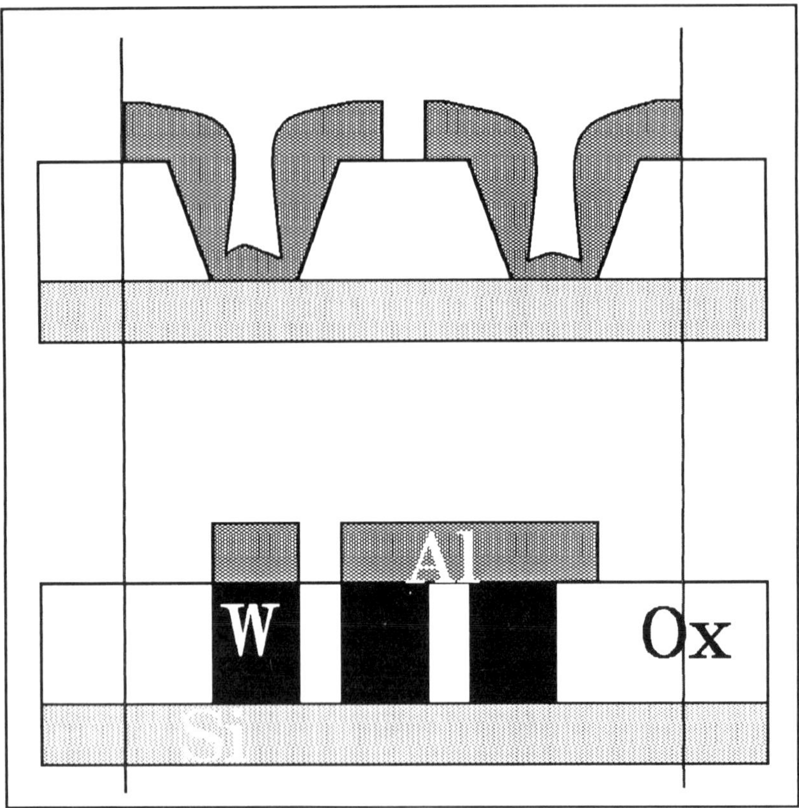

Figure 1.4. Contact with sloped walls to improve the aluminum step coverage vs. a contact filled with tungsten. Note the gain in packing density.

1.4 CONTACT PLANARIZATION AND DESIGN RULES

It is important to realize that in many designs the limit to integration is not a result of the density of the transistors and other chip components, but a result of the density of the metallization system. An often used solution is to incorporate a multi level metallization system (MLMS). In MLMS, up to four layers of aluminum, separated by dielectric layers, are incorporated to handle the needed interconnects.

As pointed out in the previous section, excellent step coverage or

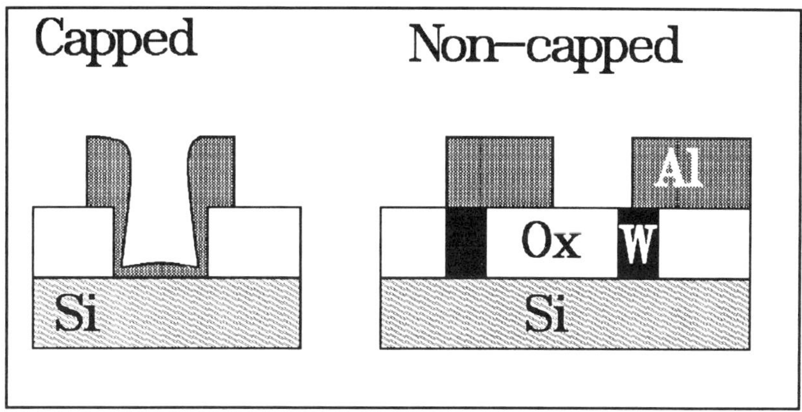

Figure 1.5.a. Cross section of a capped (left) and non-capped (right) contact.

filling of contacts and via's is required for reliability. Additionally, once the contacts and via's are filled, new (space saving) options become available. Some important advantages of fully planarized contacts/via's are:

(a) Elimination of the necessity for sloped or staircase shaped contacts/via walls. The sloped wall technique is often utilized to increase

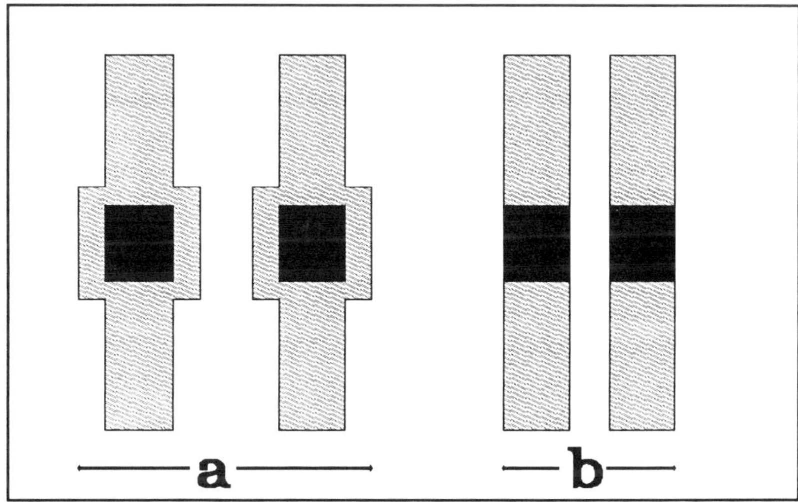

Figure 1.5.b. Top view of the situation in figure 1.5.a.: a) capped and b) non capped.

Figure 1.6. Non-stacked (left) and stacked contacts (right).

aluminum step coverage(see figure 1.4). This elimination can result in a considerable gain in density of the metal lay-out.

(b) When a good fill technique is used and the chosen fill material has good etch selectivity during Al patterning, capped (or overlapping) contacts can be eliminated (see figure 1.5.a, 1.5.b and 1.7). This provides again an increase in device density.

(c) Stacked contacts can be utilized, thus requiring less design time for the IC and giving the circuit designer greater freedom in design (see figure 1.6).

Thus, once the contacts and via's are planarized not only is there a significant improvement in device reliability, but there is also a substantial improvement in the availability of device real estate. These are the fundamental reasons behind the attractiveness of contact and via planarization as can be accomplished with CVD-W.

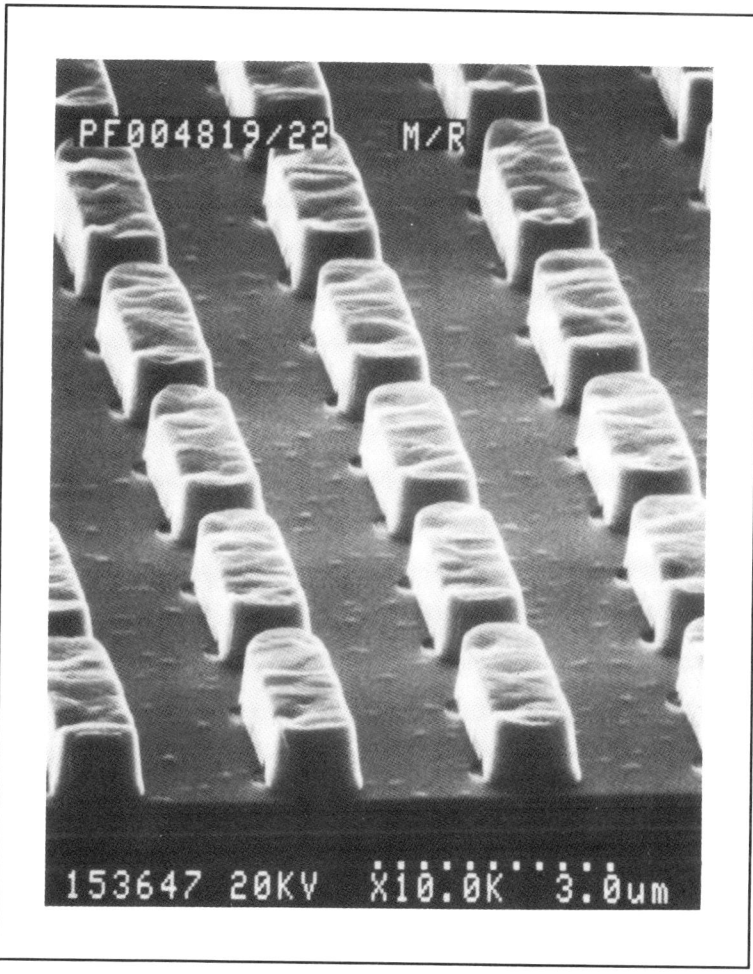

Figure 1.7. Illustration of the non-capped option of tungsten plugs. The aluminum interconnect is slightly misaligned to the right, leaving part of the plug uncovered. SEM courtesy Dr. H. Körner, Siemens AG. [from ref. 259, reprinted with permission, copyright © 1991 by Materials Research Society]

CHAPTER II

THE BLANKET TUNGSTEN APPROACH

2.1 PRINCIPAL STEPS

In this chapter we will focus on contact and via fill using the blanket tungsten approach. In chapter 5 we will discuss another application of blanket tungsten, namely, that of tungsten as the interconnect material.

Three important steps must be considered after the contact openings have been etched:

- (a) the deposition of an adhesion layer,
- (b) the blanket tungsten deposition and
- (c) the etch back of the tungsten down to the dielectric level.

The different stages are visualized in figure 2.1. Prior to the deposition of the adhesion layer the contact will under go certain preclean steps. Since these are not inherent to the blanket tungsten contact fill, we will not elaborate on the preclean steps. Blanket tungsten etch back can be regarded as the most critical and difficult step in the blanket tungsten fill process. It is therefore unfortunate that, in comparison with the tungsten deposition,

10

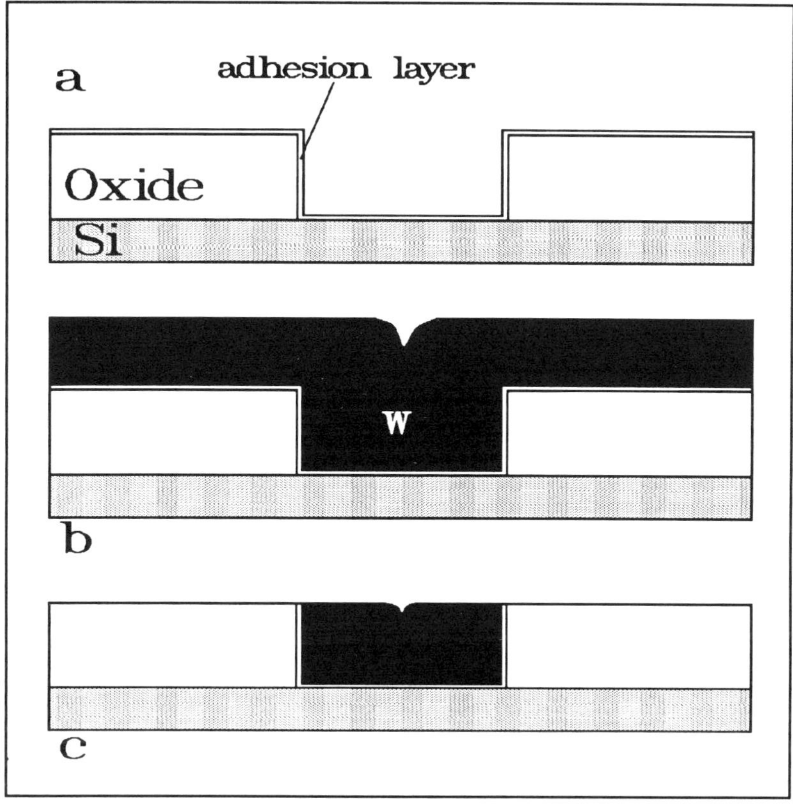

Figure 2.1. Three key steps in the blanket tungsten process: a) the deposition of the adhesion layer, b) after the blanket tungsten deposition and c) after tungsten etch back.

only a limited number of studies have been dedicated to tungsten etch back. Step (b) (and to a lesser extent step (a)), however, have received considerable attention in the literature. Many of the important results obtained will be discussed in the next paragraphs.

2.2 TUNGSTEN ADHESION

The need for an adhesion layer in the blanket tungsten process is often regarded as a disadvantage of the process. Although indeed an extra step is needed, the (sputtered) adhesion layer allows the use of the blanket tungsten fill process atop almost every substrate material, silicon, silicides, aluminum etc. This is in sharp contrast with the selective tungsten approach (see chapter III) where substrate compatibility is a major concern. It will be shown below that well known barrier layers such as TiN and TiW can also serve as excellent adhesion layers for CVD-tungsten. Since these barrier layers are already widely in use for contacts (to improve contact resistance and contact reliability) [Hoffman[1], Cohen et al.[2], Kohlhase et al.[3], Ellwanger et al.[4], Wittmer[5], Babcock et al.[6]], their use in the blanket tungsten process will not introduce additional problems with respect to contact resistance or contact reliability.

The adhesion layer must fulfill several requirements:

i) provide adequate adhesion,
ii) be compatible with the tungsten chemistry (protect the contact against phenomena as encroachment (see chapter III)),
iii) have acceptable contact resistance,
iv) have reasonable step coverage in the contact or via (about 50%),
v) provide a low defect, manufacturable process.

In the following, each of these requirements will be highlighted (see also table I and the excellent review article from Broadbent[263] and references therein.).

2.2.1 Adhesion Layers

When tungsten is deposited by means of CVD there is almost no adhesion to dielectric materials like silicon dioxide and silicon nitride. To overcome this problem an adhesion promoting layer prior to the tungsten deposition is deposited. Sputtered films such as TiW and TiN have received the most attention [Ellwanger et al.[7], Rana et al.[8]] and have proven to provide adequate adhesion. With respect to this it must be emphasized that macroscopic adhesion (Scotch tape test or bond pull test) in itself is not a valid proof of adhesion. The ultimate evidence can only be obtained when after etch back the plugs show good adhesion. In addition reliability lifetime stress tests are necessary [Kaanta et al.[142] (see also chapter III)].

Sputtered TiW and TiN are already in extensive use as barrier layers against Si diffusion in contacts. It is therefore fortunate that these layers show also good adhesion to CVD-W (i.e. no "new" adhesion material needs to be introduced). Sputtered TiN has some drawback in that, especially with the blanket H_2/WF_6 chemistry, substantial initiation times (of the order of 10 minutes) can be observed [Rana et al.[8], Iwasaki et al.[9]]. This will be exhibited by apparent lower deposition rates and thickness or uniformity control problems. The reason for the nucleation problem atop TiN is not

Table I

Properties of some sputtered and CVD adhesion layers [*]

Layer	adhesion	manufacturable	contact res.
W/Ti	ok	fair	ok
W	ok	ok	fair
Ti	weak	ok	ok
TiW	excellent	ok/fair	ok
TiW-N	ok	ok	ok
Mo	ok	ok	ok
Al	ok	ok	ok
TiN	ok	ok/fair	ok/fair
TiN (CVD)	ok	ok/fair	ok/fair
WSix (CVD)	weak	ok/excellent	ok

* see text for details

well understood at this time. Nevertheless it appears that the introduction of WF_6 and SiH_4 purge steps prior to the actual tungsten deposition can minimize the problem [Iwasaki et al.[9]]. The WF_6 reacts with the TiN under the formation of W according to:

$$2TiN \ + \ WF_6 \ \text{------>} \ W \ + \ 2TiF_3 \ + \ N_2 \qquad (2.1)$$

The reactive tungsten atoms will then catalyze in the SiH_4 purge step the decomposition of SiH_4 in hydrogen and Si. This atomic Si will initiate tungsten growth in the following H_2/WF_6 step.

Two disadvantages of this nucleation procedure appear:

(a) If the TiN in the contact is not completely continuous due to the poor step coverage (see below) of the sputter deposition, the WF_6 purge step can destroy the contact (see chapter III).

(b) The introduction of the nucleation faciliting step (not necessary in the case of TiW) gives lower wafer throughput for CVD-W on TiN versus CVD-W on TiW.

To address the issue under (a) it is more common to let the hydrogen step be preceded by a short SiH_4 reduction step (SiH_4/WF_6) which has been proven to reduce the nucleation time sufficiently in most cases.

Recently CVD-TiN has received some attention [Kurtz et al.[15], Yokoyama et al.[16], Pintchovski et al.[17], Sherman[18], Ikeda et al.[19], Buiting et al.[20], Nakanishi et al.[21], Smith et al.[22], Raaijmakers et al.[23]] and was shown to provide appropriate adhesion. The advantage of the CVD-TiN technique over the sputter deposition technique is that the step coverage of CVD-TiN can be extremely good (see 2.2.4), thus eliminating completely any problems with (a).

An alternative route to form TiN is the nitridation of sputtered Ti. This can easily be incorporated in a Ti based salicide process. An advantage is that a low and repeatable contact resistance can be obtained.

WSi_x would be a very acceptable candidate with regard to the in situ deposition possibility in the CVD-W reactor. WSi_x can relatively easily be deposited prior to tungsten deposition in the same (cold wall) reactor by using the SiH_4/WF_6 chemistry [see also chapter IX, Chiu et al.[10]]. Unfortunately, WSi_x exhibits only a moderate adhesion towards oxide which results in missing plugs after the etch back step [Ellwanger et al.[7]].

2.2.2 Chemical Resistance

It has been shown that the H_2/WF_6 chemistry, which is nowadays almost exclusively in use for blanket tungsten (see also section 2.3), can exhibit very aggressive behavior against materials such as Si, $TiSi_2$ and $CoSi_2$. This can result in encroachment and tunnel formation [Stacy et al.[11], Broadbent et al.[12], van der Putte et al.[13], Ellwanger et al.[14]] (see also chapter III) causing junction leakage. Clearly the adhesion or glue layer should not only provide adhesion but must also act as a barrier layer to protect the contact. Compatible with the H_2/WF_6 chemistry are materials such as TiW, TiN and W. Problems can be expected with Ti (formation of the non-adherent TiF_3) and Al (formation of the non conducting AlF_3) [Broadbent[263] and references therein].

When the SiH_4/WF_6 chemistry is used, the demands on the chemical compatibility of the adhesion layer are relaxed since this chemistry is so much milder than the hydrogen chemistry [Ellwanger et al.[14]]. In this way Al and Ti become acceptable adhesion layers. Unfortunately,the step coverage of the SiH_4/WF_6 chemistry is very poor (see section 2.3.2) and is therefore not suitable for contact fill applications. The silane reduction is, however, still applied to start the tungsten deposition especially atop of TiN (see 2.2.1) followed by the tungsten deposition based on the H_2/WF_6 chemistry.

2.2.3 Contact Resistance

Most of the adhesion layers shown in table I show acceptable contact resistance either to Si or $TiSi_2$. Sputtered TiN, however, has been reported to give high contact resistance to silicon. It has been claimed that by sputtering first a thin Ti layer the high contact resistance to Si can be lowered.

An advantage of the use of existing barrier layers like TiW and TiN is that the contact resistance is already characterized. The introduction of tungsten plugs in an existing process, where TiN or TiW was already in use,

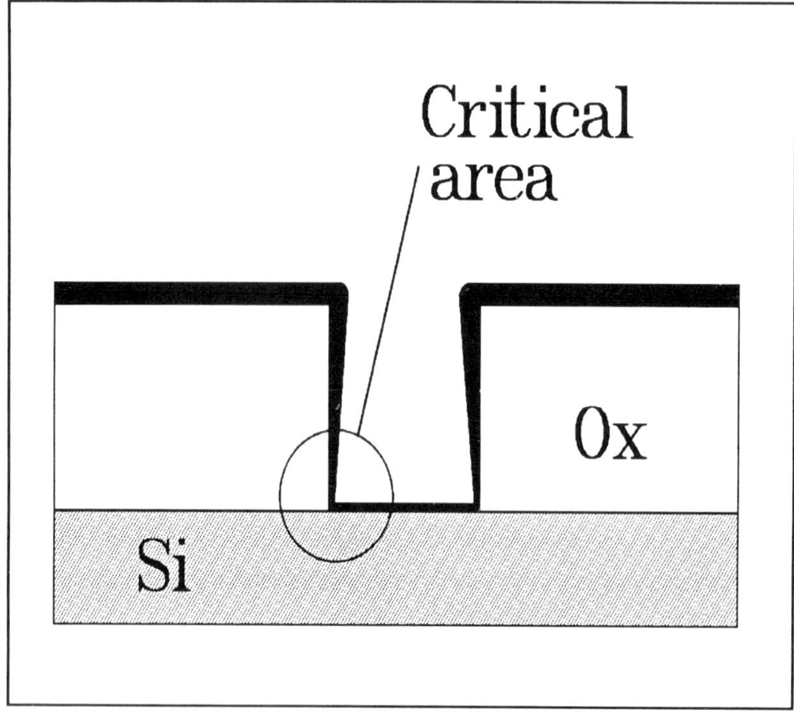

Figure 2.2. In order to keep the adhesion layer thickness in the critical areas above a certain minimum thickness the step coverage of the layer should be at least 50%.

will not change the interface which determines the magnitude of the contact resistance: the barrier layer-silicon interface. (The interface TiW/CVD-W or TiN/CVD-W is metallic and should have a very low contact resistance). Thus the electrical performance of the circuit should not be affected by any contact resistance due to the use of the plugs with the same barrier layer. The main effect of use of tungsten plugs is that the reliability of the interconnect system will improve.

It cannot be emphasized enough that the determination of the contact resistance is not a trivial matter. First, a decision about what measuring structure must be made (four terminal Kelvin, sheet end or other structures) and what correction factors for the current crowding will have to be used. Then extreme care should be taken such that no over etching of the contact down into the silicon occurs and that the correct contact size is

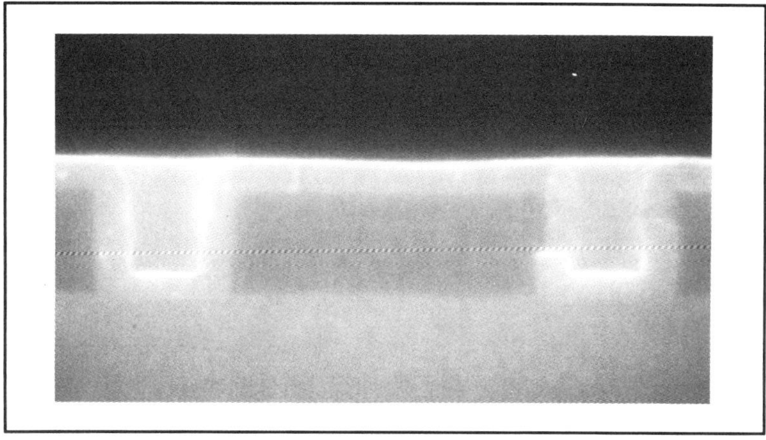

Figure 2.3. Excellent step coverage of CVD-TiN (TiCl$_4$/NH$_3$ chemistry). Contact diameter 1 μm. [SEM courtesy I. Raaijmakers, Signetics Corporation].

determined. This should be verified after the measurements by a de-processing of the samples and a careful SEM inspection. For more details see Pellogrini et al.[24], Naem et al.[25], Loh et al.[26], Scorzoni et al.[27], Wright et al. [28].

2.2.4 Step Coverage

A critical issue is the step coverage of the adhesion layer. This should be sufficient such that both the adhesion and the (chemical) barrier properties of the film are maintained. The minimum required step coverage depends upon the allowed nominal thickness at the top oxide surface (see figure 2.2) and the minimum thickness where both adhesion and the barrier properties of the material are still present. Assuming that for safety reasons a minimum thickness of the order of 0.05 μm is needed and that the nominal thickness will be of the order of 0.1 μm, then the step coverage should be 50%. For sputtered TiW in a contact of a radius of one micron and an aspect ratio of one, 50% step coverage has been shown to be achievable [Ellwanger et al.[7]].

Problems, however, are expected when the radius becomes smaller and the aspect ratio higher. Sputtering techniques will be limited here due to physical properties and CVD appears to offer the only solution. A good example could be the CVD of TiN. It has been shown that the step coverage of CVD-TiN can be excellent, see figure 2.3 [Raaijmakers et al.[23]]. A chemistry commonly in use for CVD-TiN is [Buiting et al.[20]]:

$$6TiCl_4 \ + \ 8NH_3 \ ------> \ 6TiN \ + 24HCl + N_2 \qquad (2.2)$$

The open issues of this chemistry are:

a) the high deposition temperature (ca. 600°C) needed to decrease chlorine incorporation in the film and to have acceptable growth rates. This precludes the use of this process for vias atop aluminum metallization;

b) the formation of yellow adducts of $TiCl_4$ and NH_3 on the cold parts of the reactor. This can cause particle problems which require that additional measures be taken to ensure manufacturability.

2.2.5 Manufacturability

The deposition of the adhesion layer should of course be done with a technique which does not result in high defect levels. For instance, the sputter deposition of TiW can give high particle levels although this can be resolved by proper system design. Fortunately, the sputter deposition of materials such as TiN, TiW, Ti and W is well characterized and used extensively in IC fabrication. The CVD of TiN, however, is a new technique in IC fabrication and needs further characterization before it will become accepted.

One additional consideration for making a final choice which has not yet been discussed is that of endpoint detection during etch back and etch selectivity. In the case of TiN a good etch selectivity of CVD-W over TiN (say 15:1) is obtained with commonly used gases. As well, TiN provides an easily detectable nitrogen signal for optical endpoint detection.

Another point of concern is that many sputter systems use some type of clamping to keep the wafer in place. Thus, some areas at the edge of the wafer will not have an adhesion film. The result will be tungsten peeling in these areas during the subsequent blanket tungsten deposition. One way to overcome the problem is to use clamps in the tungsten deposition system which are designed such that they will prevent tungsten deposition on the edge of the wafer (see chapter VII).

Normally the backside of the wafers doesn't have an adhesion layer but consists of either (poly crystalline) silicon or some type of oxide. When no special precautions in the CVD-reactor are taken. tungsten will also deposit on the back side of the wafer. Especially in the case of an oxide layer at the back side, the adhesion will be very poor. During further processing of the wafers, the tungsten starts to peel. The result is a tremendous particle problem in the IC production line. Even in the case where there is silicon on the backside of the wafer often peeling will occur. An early fix of the problem was to deposit the adhesion layer also on the backside. This, of course, implied several additional process steps and was therefore only considered as temporarily. Nowadays, the tungsten deposition equipment solves the problem by shielding, backside purge or in other ways (see chapter VII).

2.3 BLANKET DEPOSITION OF TUNGSTEN

In the next sections we will discuss the fundamentals of the non-selective deposition of tungsten. Much attention is paid to the phenomena of step coverage since this is a key issue for successful filling of contacts.

2.3.1 Chemistry

Many chemistries are available to deposit tungsten (see chapter VI). Here we will only discuss the most frequently used ones: SiH_4/WF_6 and H_2/WF_6. In these chemistries silane and hydrogen act as reducing agents for the tungsten source, respectively. The overall equations can be written as:

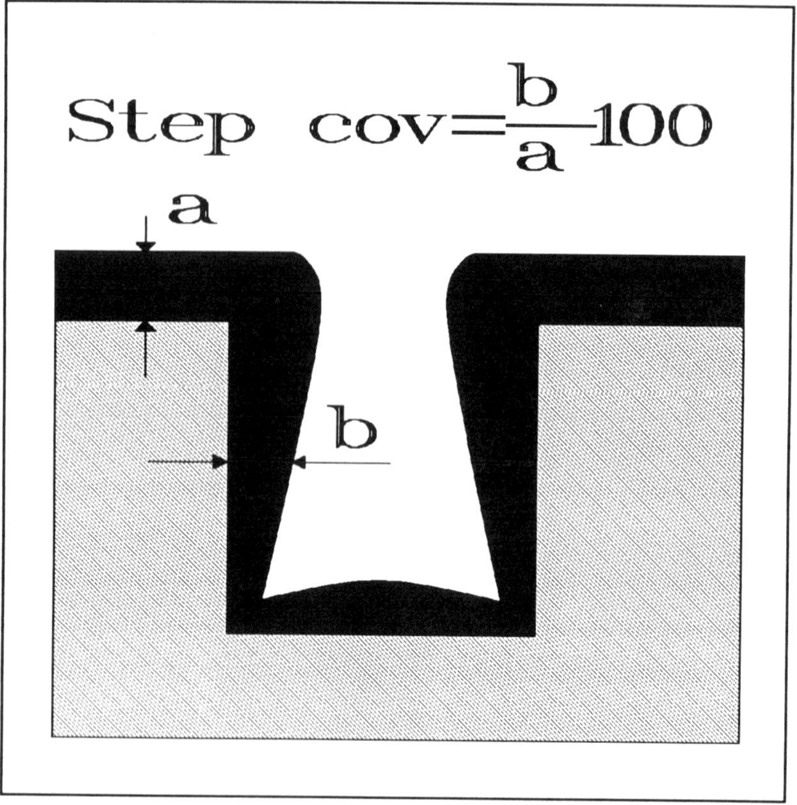

Figure 2.4. Definition of the step coverage in this book. The side wall thickness b is measured at half depth of the trench or the contact.

$$WF_6 + 3H_2 <======> W + 6HF \qquad (2.3)$$

and:

$$2WF_6 + 3SiH_4 <=====> 2W + 3SiF_4 + 6H_2 \qquad (2.4)$$

Equation 2.4 is in conflict with thermodynamic prediction since hydrogen (formed in reaction 2.4) normally will react with WF_6 to form HF (according to reaction 2.3) but has been experimentally proven to be correct [Yu et al.[29], Sivaram et al.[30]] (see also chapter VI). This implies that the silane chemistry proceeds far from equilibrium [see also section 3.4.2]: the formation of HF via reaction 2.3 is kinetically blocked, i.e., slow in comparison with the SiH_4 reduction.

2.3.2 Step Coverage

Let us first define step coverage (see figure 2.4) as the ratio of the thickness of the tungsten film at the side wall at half depth and the nominal tungsten film thickness. This is a purely arbitrary definition but has been proven to work in practice. In an ideal case we want to have 100% step coverage, that is the growth rate at each surface is equal.

Table II

Step coverage in deep trenches

H_2		SiH_4		WF_6		Step cov.	Dep. Rate
sccm	Torr	sccm	Torr	sccm	Torr	%	A/min
2000	0.17	0	0	100	0.009	80	250
1940	0.17	60	0.005	100	0.009	53	588
1900	0.17	100	0.009	100	0.009	25	1170

Data from ref. 31 Chuck temp.$=430^{\circ}C$, $P_{tot}=200$ mTorr

Why is the step coverage of such an importance? (See figure 2.5). If the step coverage is lower than 100% there will be a cavity (or void) formed during the deposition; a so-called key hole. The size of this key hole depends on the magnitude of the step coverage. During the etch back of the tungsten layer such a key hole can cause destruction of the contact (see figure 2.5.).

In this paragraph we will give more insight into the parameters which determine step coverage in tungsten CVD but first we will give some experimental facts.

The experimental facts: In an exertion to find the best chemistry for contact fill it was found [Schmitz et al.[31], Blumenthal et al.[31]] that there exist substantial differences between the SiH_4/WF_6 , $H_2/SiH_4/WF_6$ and H_2/WF_6 chemistries. The data was obtained using deep trenches (more than 10 um deep and about 2 um wide) as the step coverage monitor (see figure 2.6). It has been shown both experimentally [Schmitz et al.[31], Hasper et al.[32]] and with simulations [Hasper et al.[32]] that there is a one to one correlation (see below) between the step coverage in contacts and those obtained in trenches (see also figure 2.15). The use of trenches to study the step coverage of CVD films was introduced by Levin and Evans-Lutterodt[33] for oxides. The advantage of trenches over real contacts as a step coverage monitor is that a cross section for SEM inspection is much easier to obtain. If the cross section in the case of a contact is not made exactly through its center, incorrect step coverage data may be obtained.

The effect of various chemistries on step coverage are shown in table II. From such data one can determine that in order to obtain optimal step coverage the H_2/WF_6 chemistry is the appropriate choice. In the next sections we will discuss some theoretical backgrounds of mechanisms which influence the step coverage of a CVD process and explain why the H_2/WF_6 chemistry gives better step coverage than the SiH_4/WF_6 chemistry.

The physical approach: One approach is to take advantage of the fact that under low pressure conditions the mean free path length (L) is much longer than the typical dimensions of the contact or via holes. L can be calculated using expression 2.5 [Dushman[261]]:

$$L = \frac{1}{2^{1/2} \, \pi \, n \, d^2} \tag{2.5}$$

where n is the concentration in molecules/cm^3 and d is the molecular radius. For the case of hydrogen and WF_6 this expression gives at 200 mTorr L equal to about 200 and 100 μm, respectively. Since the contact size will be of the order of 1 um we can consider the deposition as a truly "line of sight" behavior. This implies that the deposition rate at a given point at the surface is proportional with its solid angle α (assuming a high sticking coefficient).

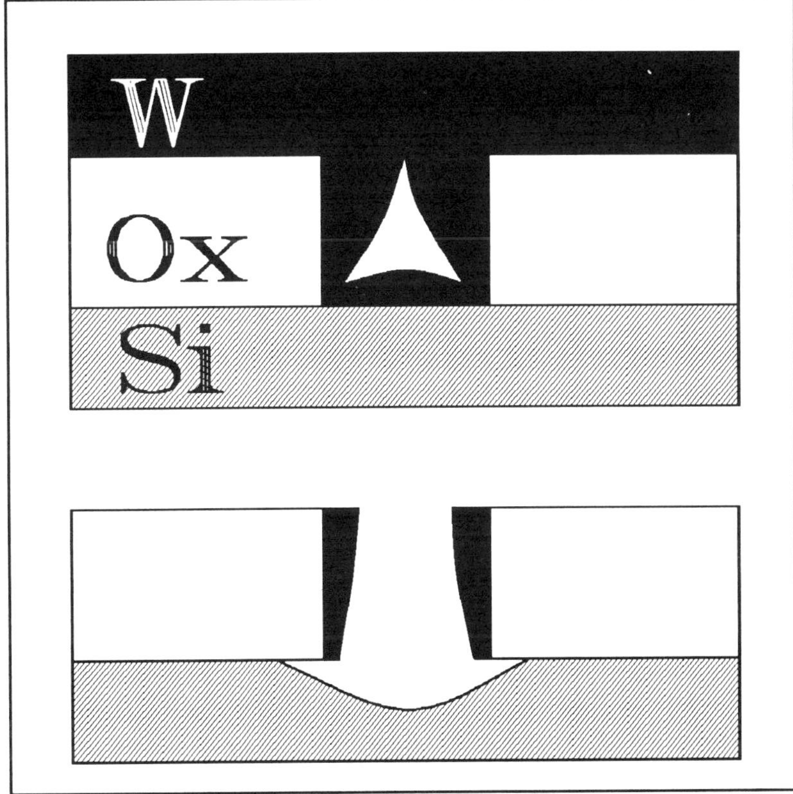

Figure 2.5. Formation of a key hole in case of insufficient step coverage (top) and the situation after etch back (bottom).

Such an approach is usually followed for physical vapor deposition (PVD) (i.e. sputter deposition or e-beam evaporation). In figure 2.7 we have given the situation for a two dimensional case. Inside the contact the deposition rate can be described by:

$$\text{dep. rate} = \text{constant} \cdot \alpha \tag{2.6}$$

where α is the solid angle for any point at the surface. After a short time of deposition the film profile will have the shape as depicted in figure 2.7.b and we note that we would always end with the formation of a keyhole.

The description above is only accurate if the sticking coefficient is very high i.e. at the first collision with the surface the molecule will adsorb

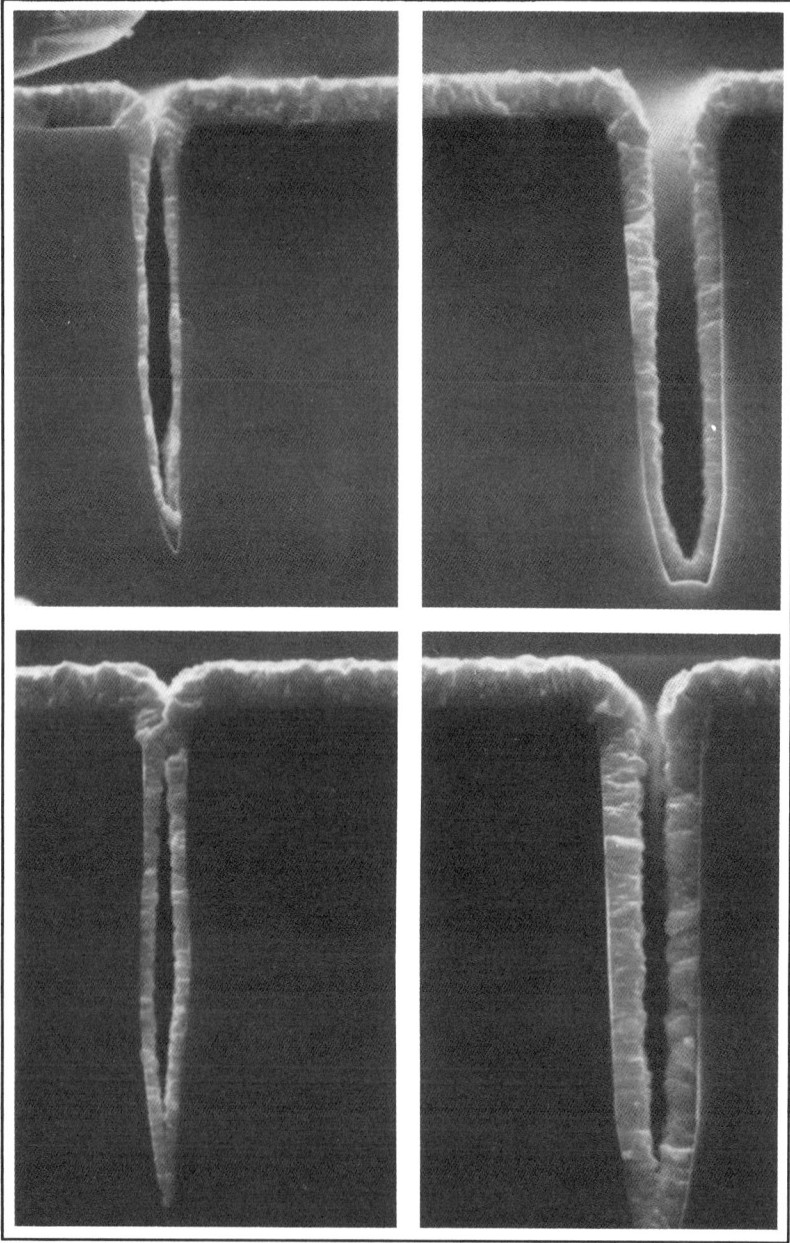

Figure 2.6. Step coverage of CVD-W in deep trenches. SiH_4/WF_6 chemistry (top) and H_2/WF_6 chemistry (bottom), both at 413°C and 820 mTorr. Depth is about 10 μm.

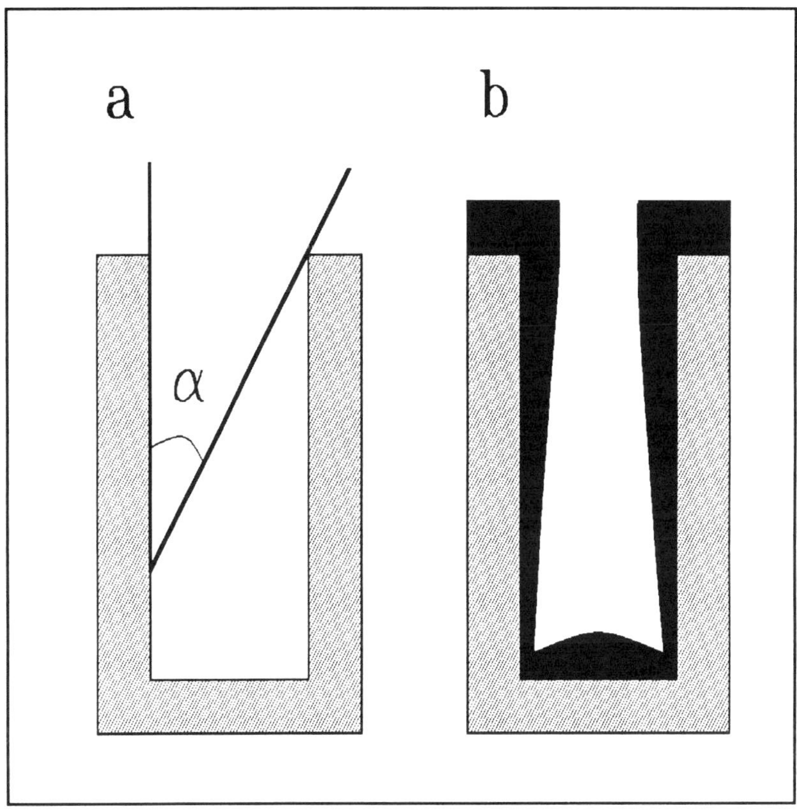

Figure 2.7. Two dimensional representation of a line of sight deposition (a) and the corresponding film profile (b).

and will not return to the gas phase. This appears in general not to be true in LPCVD-W. A simple calculation can illustrate this. At a partial WF_6 pressure of 10 mTorr, the impingement rate of the WF_6 molecules is 9.4×10^{20} cm^{-2} min^{-1}. When every molecule would react to form tungsten, the resulting growth rate would be 2.4×10^5 nm/min. Under typical conditions, the growth rate is only 30 nm/min for reaction 2.3 and about 100 nm/min for reaction 2.4. From this we conclude that the "chemical" sticking coefficient is extremely low (about 0.0001). Almost every molecule leaves the surface without further reaction. Under such conditions equation 2.6 is no longer valid.

The situation can be further illustrated by the following. Assume a

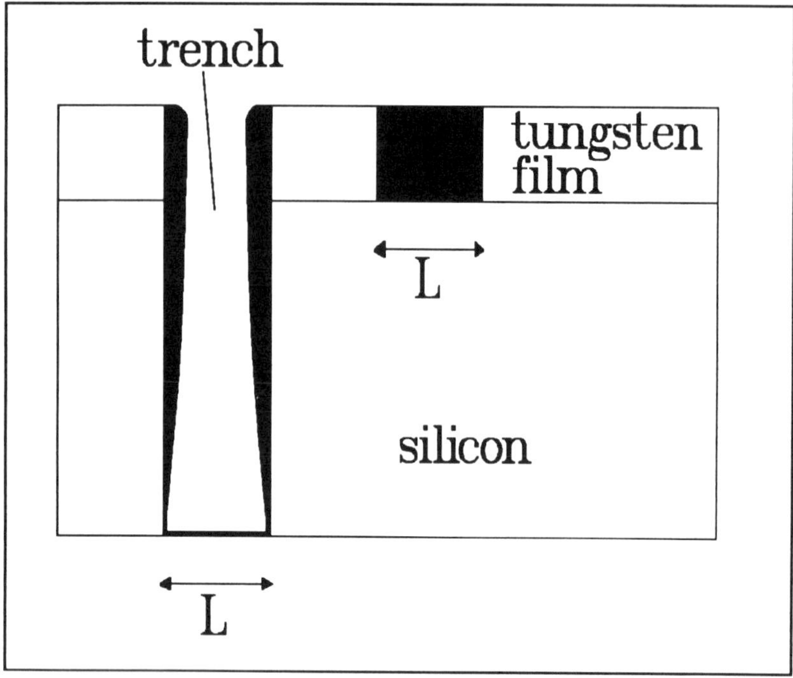

Figure 2.8. The cross sectional area of tungsten in the trench should approximately be equal to the black area on the right. Assume a sticking coefficient of 1 and no surface diffusion.

sticking coefficient of one and no further redistribution of material due to surface diffusion. Under these circumstances we can calculate what the maximum amount of tungsten we can expect in a trench (see figure 2.8). This amount is roughly equal to the amount of tungsten which would otherwise have been deposited on the surface area now spanned by the trench opening. In fact this has been investigated for hydrogen and silane chemistries [Schmitz et al.[31]] in trenches 10 um deep and about 2.5 um wide. It appears that about 90% of the total amount of tungsten which was deposited in the trench was due to redistribution. This high value together with the extreme low sticking coefficient points clearly in the direction that tungsten deposition by CVD cannot be adequately described by a simple line of sight approximation.

The calculation of the step coverage using a line of sight model and assuming a low sticking coefficient is not a trivial task. There have been attempts in the literature [Tsai et al.[34], Shaw et al.[35], Yuuki et al.[36], Cheng et al.[37], Okada et al.[38]] to solve the problem using a Monte Carlo type of simulation. However, assumptions about the sticking coefficient and thus the identity of the absorbed species have to be made. The predicting power (in terms of process optimization) of such an approach is not as high as the method which will be described below. Nevertheless, the Monte Carlo method can give more insight into what actually happens at the surface in terms of molecular reaction steps, surface diffusion and sticking coefficient.

The chemical approach: Fortunately, there is another method available to predict the step coverage of CVD-W. We use the fact that there is a close similarity between the filling of a contact hole and the transport mechanism operative in heterogeneous porous catalysis. This connection was made for the first time by McConica [McConica et al.[39], Chatterjee et al.[40]] and below we roughly follow their approach. Consider figure 2.9.

We have to realize that the growth rate at any time at any given surface point is determined by the reaction kinetics. It has been found [Broadbent et al.[44], Pauleau et al.[45], McConica et al.[46]] that for the H_2 reduction the deposition rate of tungsten can be described by:

$$\text{Rate} = \text{constant x } e^{-Ea/RT} \text{ x } [P_{WF6}]^0 \text{ x } [P_{H2}]^{1/2} \qquad (2.7)$$

where P_{WF6} and P_{H2} are the partial pressures of WF_6 and H_2 respectively. For the silane chemistry it has been found [Schmitz et al.[48], Rosler et al.[49]] that the deposition rate of tungsten can be described by:

$$\text{Rate} = \text{constant x } [P_{WF6}]^0 \text{ x } [P_{SiH4}]^1 \qquad (2.8)$$

where P_{SiH4} represents the silane partial pressure. We see that the local pressures of the reducers determine the local growth rates and consequently the step coverage. However, there is one limitation to this model, when the concentration of WF_6 becomes very small ("zero") such that the growth rate drops to zero, equations 2.7 [Kleijn et al. [163]] and 2.8 are no longer valid.

The local concentrations needed in these equations are determined by the rate of consumption (because of the surface reactions) and the rate

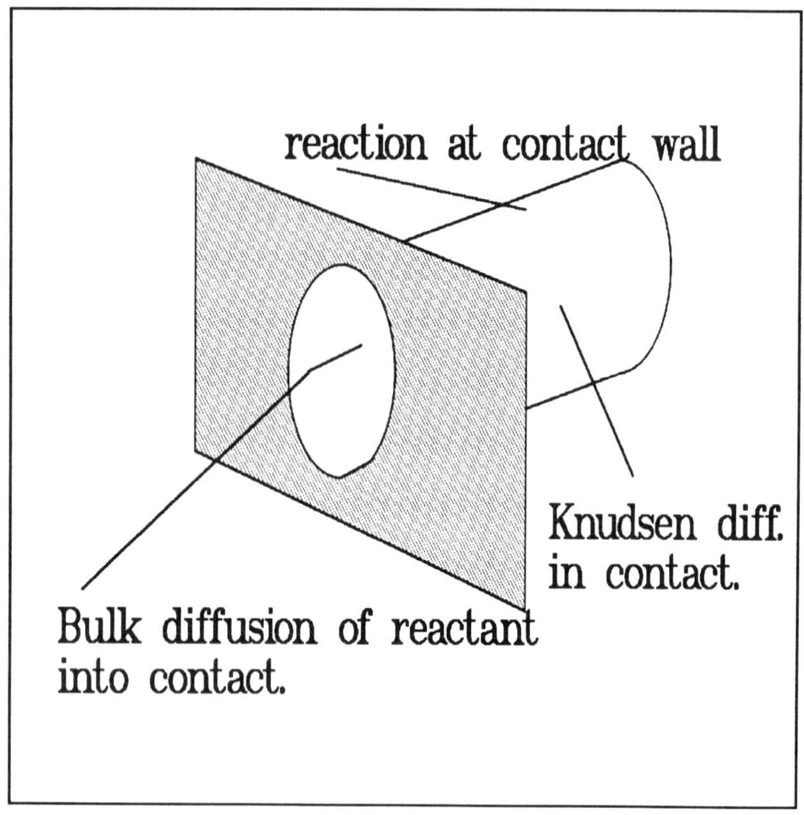

Figure 2.9. Mass transport in a small pore (or contact).

of diffusion into the contact. Under low pressure conditions there are two diffusion transport mechanisms operating. First, we have transport of the reactants from the bulk in the reactor to the wafer surface. Since the mean free path in the reactor under typical LPCVD pressures is much shorter than the reactor dimensions, we deal here with bulk diffusion transport. The bulk diffusion constant can be described by the Einstein equation:

$$D_{bulk} = 1/3 \; L \; v \qquad (2.9)$$

where v is the thermal Maxwell velocity. Second we have diffusion of the reactants into the contact hole. As noticed above, however, the mean free path is much longer than the size of the opening and consequently the diffusion is of the molecular or Knudsen type. The magnitude of the diffusion coefficient is now given by:

$$D_{Knudsen} = 2/3 \; r \; v \qquad\qquad (2.10)$$

where r is the radius of the contact. Since at low pressures L can be in the range of 100 to 1000 um and r is of the order of 0.5 um, we see that there can easily be two or three orders of magnitude difference in the bulk and Knudsen diffusion rates. Thus the mass transport of both reactants and products inside the contact hole is much slower than in the bulk of the reactor and concentration gradients can easily develop. As soon as concentration gradients develop (especially of the reductors) there will be a degradation of the step coverage.

Equation 2.10 shows two more important features of the mass transport in the contact:

i) The rate of the diffusion is pressure independent as opposed to the bulk diffusion rate (note that the free path way L is inversely proportional to the pressure). This is true as long the pressure is low (<100 Torr). In this pressure range the diffusion in the contact will be slower than that in the bulk of the reactor. Above 100 Torr we have bulk diffusion in both the reactor and the contact, Since the rate of diffusion at these pressures is at least two orders of magnitude slower than at 1 Torr pressure, the reactor tends to run in a diffusion controlled regime. This will degrade the step coverage (see also the discussion under high pressure processing and figure 2.13).

ii) During tungsten growth the diameter of the contact will decrease and as a result the diffusion slows down. Although the step coverage at the onset of the process can still be excellent it degrades during film growth. From this we can conclude that it does not make much sense to give step coverage figures without defining exactly in what geometry and at what nominal film thickness they are obtained.

Bearing in mind the above discussion, we can now better understand the lower step coverage of the silane chemistry as compared to the hydrogen chemistry. Three key factors are:

i) The diffusion coefficient of silane is about 4 times smaller than that of hydrogen. Thus concentration profiles will emerge much earlier for silane than for hydrogen leading to a lower local growth rate (and consequently lower step coverage).

ii) The reaction order for silane is one (equation 2.8) whereas that of hydrogen is only 1/2. Thus in the event of reactant starvation, the effect on growth rate is larger in the silane case than in the hydrogen case.

iii) The heterogenous rate constant is larger for silane than for hydrogen giving more constraints for mass transport to keep in line with mass consumption.

Calculation of step coverage: Step coverage can now be calculated using the equations of Fick and equation 2.7 or 2.8. Because of the mathematical complexity, no analytical equations can be obtained and numerical methods are necessary (see below). However, when we make some simplifying assumptions for the hydrogen case an equation can be obtained which has proven to give correct practical predictions [McConica et al.[39]]. The assumptions are:

i) no growth at the bottom of the contact,

ii) no hydrogen concentration profile inside the contact (because of the high diffusivity of hydrogen),

iii) equation 2.9 also valid for very low WF_6 concentrations.

The interested reader will recognize that all these assumptions will cause an overestimate of the actual step coverage.

The result is (for contact holes):

$$\text{Step cov.} = 1 - A(P_{H2}^{1/2} / P_{WF6})^{1/2} \tag{2.11}$$

with

$$A = [h/r] \times [0.5 \exp^{-8800/T} \sqrt{(M/T)} 10^{-4}]^{1/2}$$

where h is the depth of the contact, r the radius, M the mass of WF_6, and T the absolute temperature. Note that A is temperature dependent (increases with temperature). At 400^0C, $P_{H2} = 0.2$ Torr, $P_{WF6} = 0.02$ Torr, depth of contact 1 um and a radius of 0.5 um, equation 2.11 yields 28% step coverage. We will see in the next section that equation 2.11 will give valuable information for process optimization in terms of step coverage.

The benefit of high pressure processing: Equation 2.11 can be used to develop a process with superior properties than the processes in table II. To start with, we observe that although the hydrogen based process has relatively good step coverage, the growth rate is rather low (25 nm/min.). Assume we need a layer thickness of 800 nm. The process time for this film thickness will be at least 32 minutes not counting the load/unload and process overhead time (purge steps, pump down etc.). This is totally unacceptable for a single wafer system and even for a batch type of system the situation is not very favorable. In addition, it appears that for high aspect ratio contacts (>2) the step coverage is still not sufficient with this process.

When we consider equation 2.7 again we notice that there are two ways to increase the growth rate: via the temperature and/or the hydrogen pressure. From a manufacturing standpoint we want to have identical processes for both contact and via fill (can run in the same reactor using the same process). For via fill, however, the maximum allowable wafer temperature will be about 400^oC with regard to the underlying aluminum metallization. Thus the only parameter left is the hydrogen pressure. However, when we increase *only* the hydrogen pressure we see that the step coverage will degrade since the factor $[P_{H2}^{1/2}/P_{WF6}]$ increases and as a result the step coverage decreases. A simple way to overcome this is to increase the *total* pressure of the system by throttling the pump speed. In this case both the growth rate will increase, because of the rise in P_{H2}, and the step coverage, because of a decrease in $[P_{H2}^{1/2}/P_{WF6}]$.

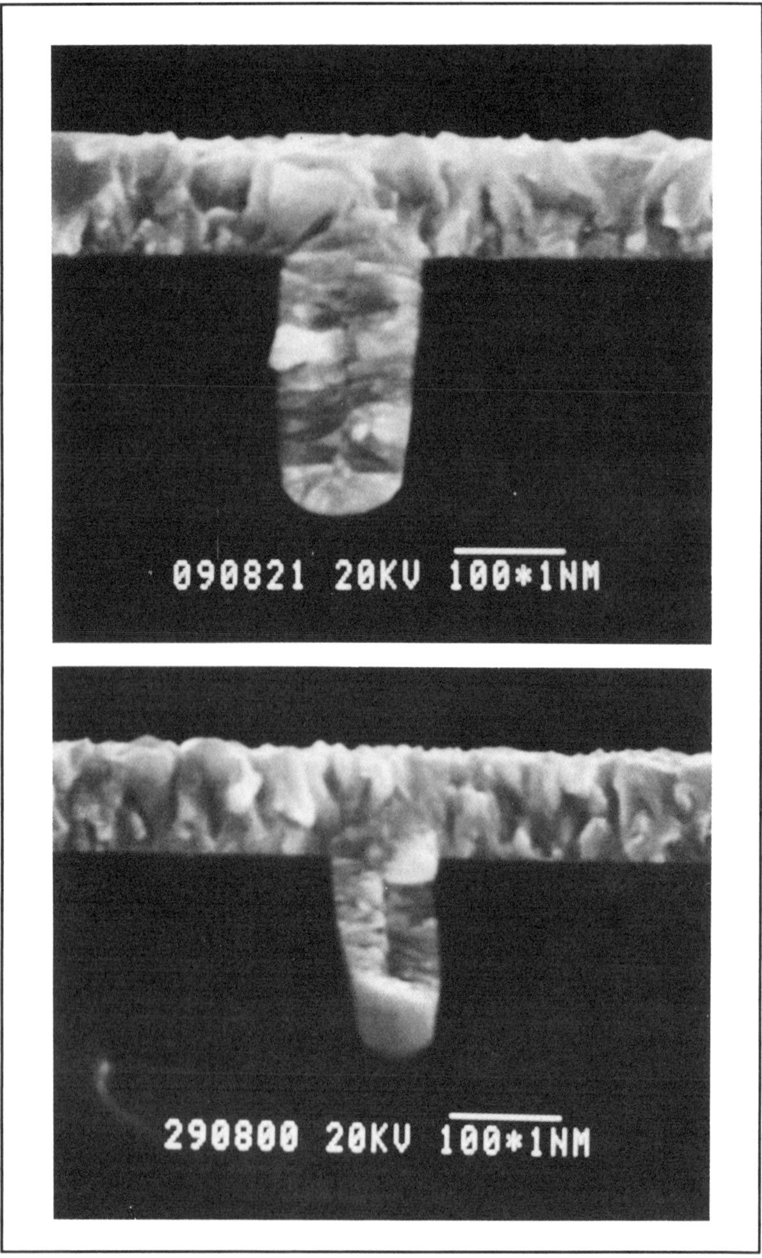

Figure 2.10. Some results of high pressure processing. Contacts filled at 30 Torr, 400°C (top) and at 90 Torr, 480°C. [SEM courtesy of S. Kang, Genus, Inc.]

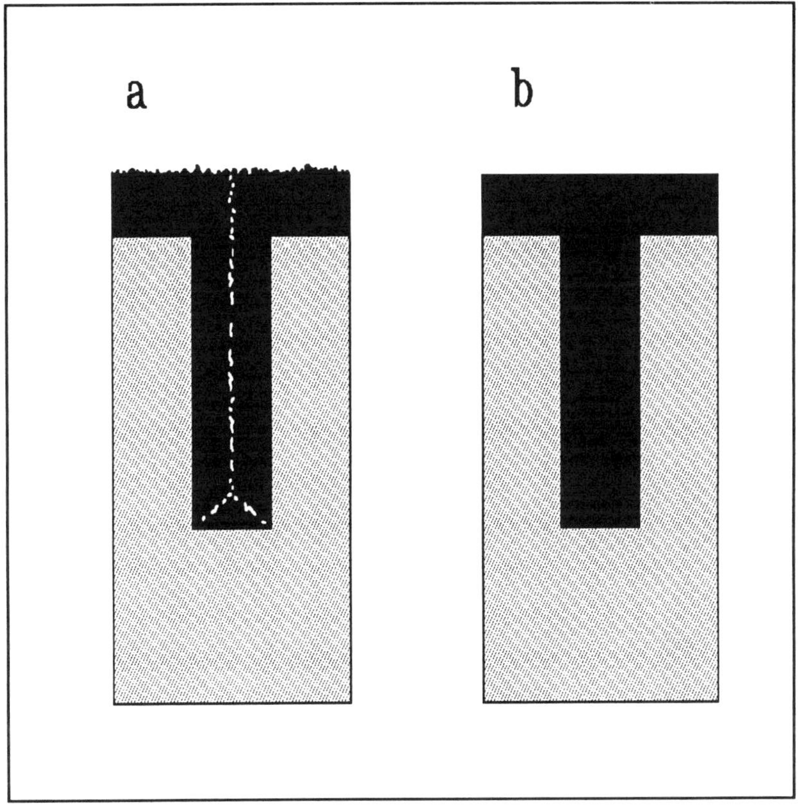

Figure 2.11. Formation of the seam at low (a) and high (b) pressure processes.

In doing this, however, we have the risk of running into a WF_6 starvation problem which often gives bad thickness uniformities, lower growth rates, and a degradation of the step coverage. A simple increase in the WF_6 flow is sufficient to prevent this problem. Schmitz et. al.[50], were the first who pointed out the benefits (i.e. not only improved step coverage and throughput but also better control of wafer temperature, see chapter VII) of going to higher process pressure (up to 12 Torr). Some results of high pressure processes are shown in figure 2.10. Joshi et. al.[51] and Clark et. al.[52] confirmed these results and expanded the pressure range up to 80 Torr.

By going to higher pressures one other advantage appears, namely, the roughness of the film decreases. In principle this gives a better closure of the contact (minimum seam without micro voids present as is the case at lower pressures) (see figure 2.11). The result is that after etch back the dimple in the center of the plug (characteristic for the blanket process, see figure 2.12) can be much smaller. See chapter V for more details about film roughness.

Is there an upper limit for the pressure with respect to the step coverage? The answer depends on what approach is taken. McInerney et. al.[53] showed that depending on the deposition temperature there is a restricted pressure window where the step coverage is 100% (see figure 2.13). At low pressure the step coverage is not optimal because the WF_6 pressure at the contact inlet is not high enough to prevent the WF_6 "zero" pressure condition in the contact during late growth stages. At pressures that are too high, the bulk diffusion slows down such that starvation at the wafer surface can occur leading to low WF_6 (contact) inlet pressure. At 450°C the 100% step coverage window is from about 0.5 to 10 Torr. This window can be enlarged to some extent by increasing the WF_6 partial pressure (at the expense of WF_6 utilization) Recently, it was shown by Bartholomew et al.[54] that even under atmospheric conditions excellent step coverage can be obtained (see figure 2.14). Two important features were used in that study:

i) A very high gas velocity was employed in the reactor such that essentially there is no concentration gradient development in the gas phase (Peclet number much larger than one) and the WF_6 wafer surface concentration is close to the reactor inlet WF_6 concentration.

ii) Very high WF_6 partial pressures were used (10 to 50 times higher than usual). This will prevent the occurrence of the zero WF_6 concentration condition in the contact during growth.

Numerical calculation of step coverage: Although equation 2.7 is very useful it is only qualitative and more over, is only valid for the hydrogen chemistry. Some attempts have been made to calculate step coverage more accurately by using numerical solutions to the diffusion equations. Hasper et al.[32] found that, in order to have a reasonable agreement between the calculations and the experimental data, thermal diffusion (see also chapter VII) had to be included in the model. They were able to show a close

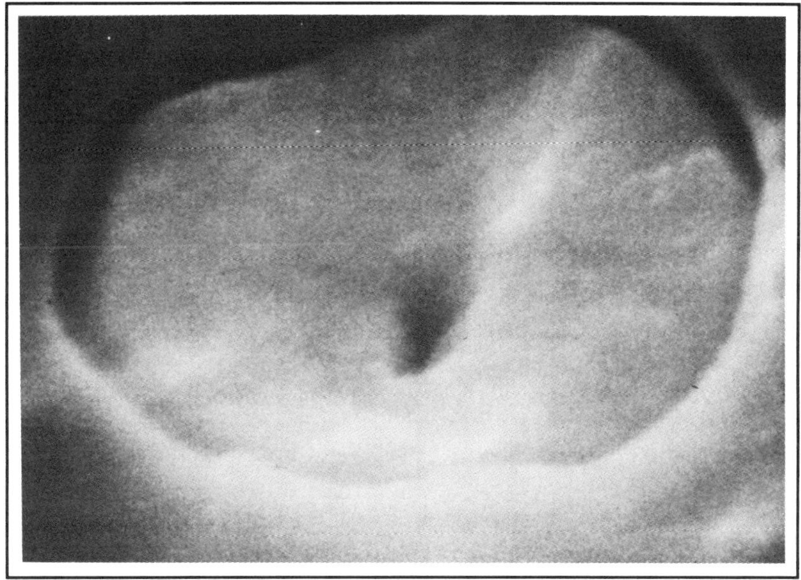

Figure 2.12. Example of the dimple in the plug after etch back (low pressure deposition process).

correlation between the step coverage in trenches and contact holes which was earlier suggested by Schmitz et al.[31] (see also figure 2.15 and 2.16).

The concentration profiles as calculated for the reactants and products of the silane chemistries at different stages of the tungsten film growth are depicted in figure 2.17 [Schmitz et al.[43]].

The Thiele modulus: In designing reaction conditions for optimal step coverage, it appears that a dimensionless number, the Thiele modulus (N_T), is very useful. This number describes whether a reaction in a porous catalysts will be mass transport limited. Translated into terms of step coverage this means that as soon as the deposition inside the trench becomes diffusion controlled concentration gradients will develop and consequently the step coverage tends to degrade. It has been found that if N_T^2 is larger than 0.25 the reaction can be considered as being mass transport limited.

For a zero order (WF_6 is zero order and the limiting species!!)

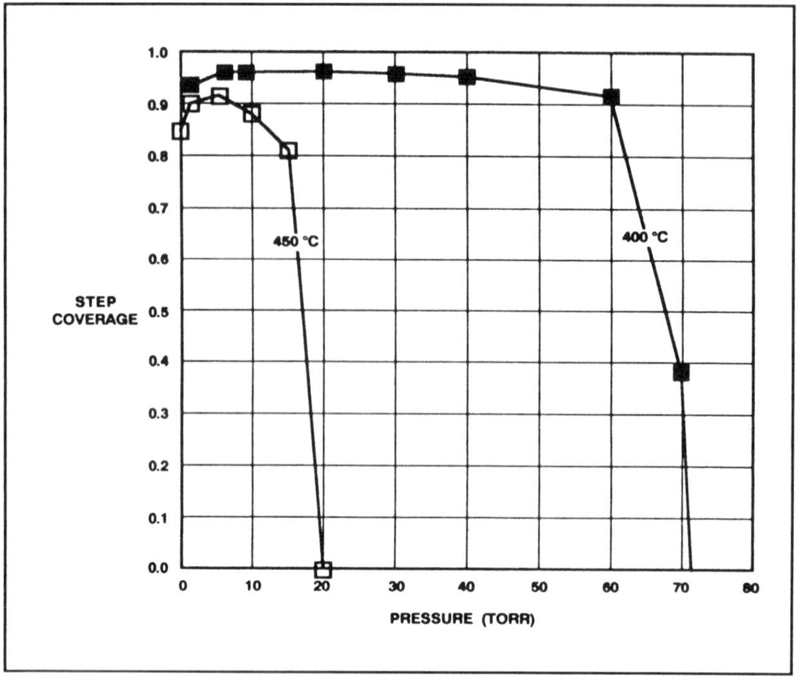

Figure 2.13. For a given temperature there is only a limited pressure window where the step coverage is 100%. [From McInerney et al.[53] reprinted with permission]

reaction and, the depth twice that of the radius of the contact, N_T becomes:

$$N_T^2 = 2 \, x[Rxd]/[DxC] \tag{2.12}$$

For typical conditions of the H_2/WF_6 reaction at the **onset** of deposition [at 200 mTorr and 400°C] N_{T2} becomes :

$$N_T^2 \approx 10^{-6}$$

with the parameters:

$$R = 300 \text{ A/min} = 1.4x10^{-11} \text{ mol/cm}^2 \text{ sec}$$
$$d = 1 \text{ um } (=\text{depth of the contact})$$
$$D = 1.5 \text{ cm}^2/\text{sec } (= \text{Knudsen diff. coefficient})$$

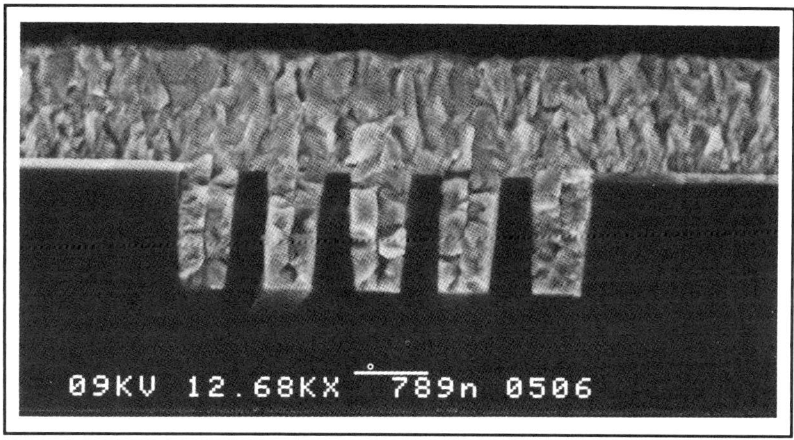

Figure 2.14. Example of high aspect ratio trenches filled with CVD-W under atmospheric conditions. [SEM courtesy of L. Bartholomew, Watkins Johnson Company, adapted from ref 54].

$$C = 1.2 \times 10^{-9} \text{ mol/cm}^3 \text{ (WF}_6 \text{ concentration)}$$

which is much smaller than 0.25. Indeed for this deposition, good step coverage is found [Schmitz et al.[31]]. In fact one needs to repeat the calculation for several stages of the fill process. The end stage of the fill process is of course the most critical one since the Knudsen diffusion coefficient becomes smaller due to the decreasing contact radius (see equation 2.10). One should bear in mind that a small Thiele modulus does not always ensures good step coverage. This also depends on the reactor mode ie. surface or diffusion controlled deposition rates.

Surface diffusion: Good step coverage with CVD reactions as described in the literature often is related to the phenomenon of surface diffusion. When a gas is absorbed at a surface it can behave like a two dimensional gas. The concept is used in the field of heterogenous catalysis in which exists indications that the mechanism can contribute substantially to total mass transport [Barrer[42], Dacey[41]]. Hence, in CVD of tungsten it is in principle possible that an absorbed molecule of WF_x is very mobile and can counteract the concentration gradient built up in the gas phase. In this way a higher step coverage will be obtained than in the case there would be no surface diffusion. A recent study to unravel this, however, found that the

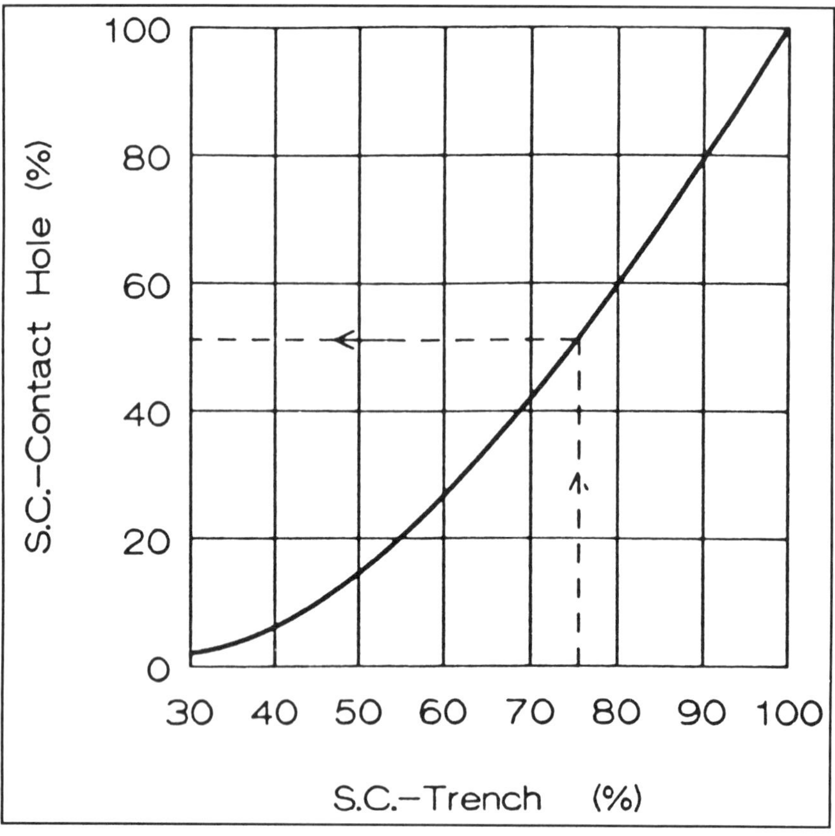

Figure 2.15. Correlation between the calculated step coverage of trenches and of contacts/via holes. [From Hasper et al.[32] Reprinted with permission].

amount of surface diffusion in CVD-W is negligibly small and of no importance [Schmitz et al.[43]]. More support for this view point can be obtained from the simulation by Hasper et. al.[32]. They showed good agreement between calculated and experimental step coverages (for the hydrogen case) using a model with no surface diffusion incorporated.

2.3.3 Film Thickness Requirements

Assuming 100% step coverage, complete closure of the contact

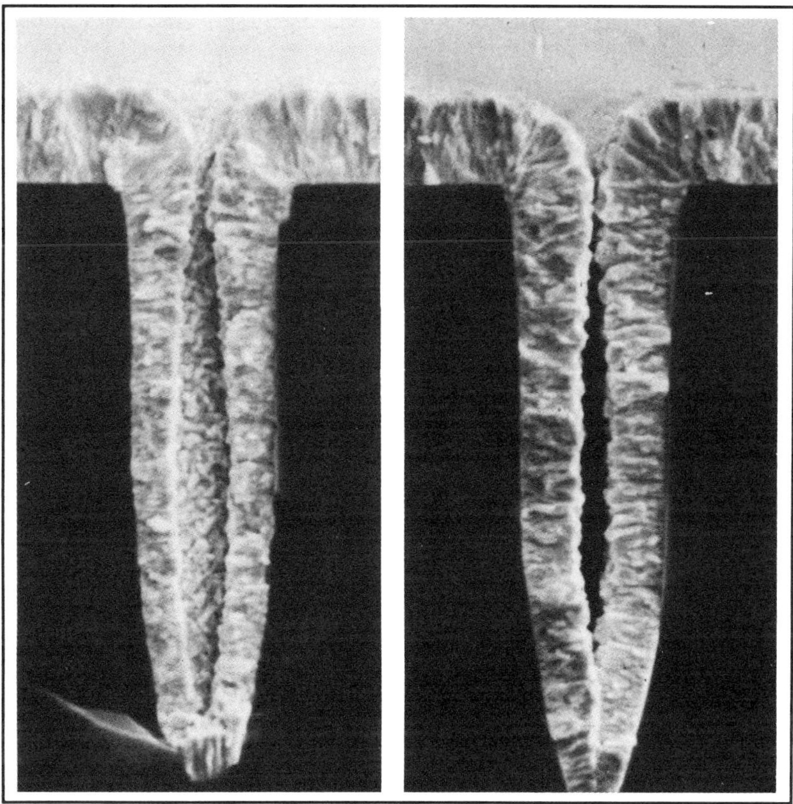

Figure 2.16. Illustration of figure 2.15. Contact (left) and trench (right) with same aspect ratio (about 3). Opening about 2 μm. [SEM courtesy of A. Hasper].

should occur when the film thickness becomes identical to the radius of the contact or via. This has been experimentally explored. For contacts with a diameter of one micron, 600 nm of tungsten were needed [Schmitz et al.[31]]to close the contact. After etch back, acceptable plugs with almost no dimple (see below) were found. Due to the non uniformity in both the deposition thickness and the etch back rate, however, in practice a thicker film is needed. A typical thickness is 800nm for 0.9um contacts (see below).

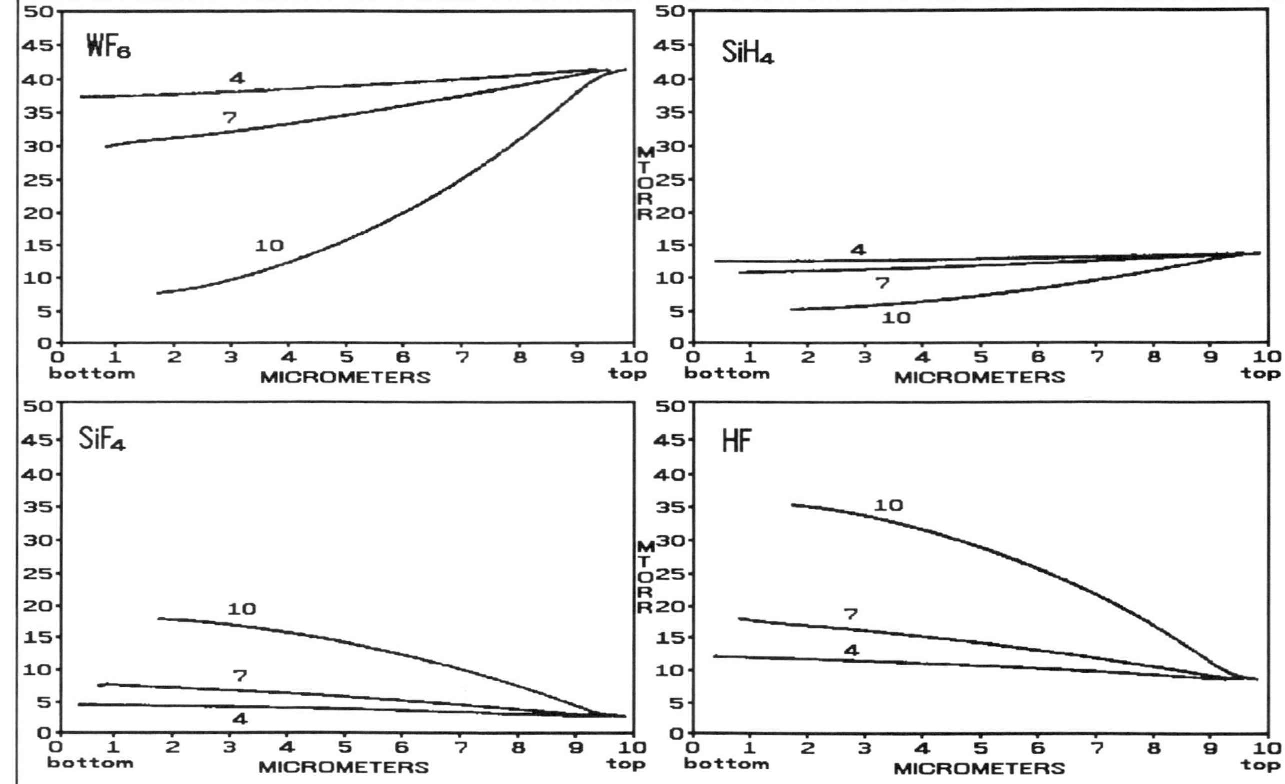

Figure 2.17. Concentration profiles at 4, 7 and 10 minutes tungsten growth in 10 μm deep trenches. SiH_4/WF_6 chemistry. [From reference 43, reprinted with permission].

2.2.4 Film Thickness Uniformity

It is not possible to state in general what the allowable film thickness non-uniformity should be. The most important input parameter here is the question, "what remaining aspect ratio (due to the etching out of the plugs) is allowed after etch back". In other words, how deep below the oxide level are the plugs after etch back? Due to the thickness non-uniformity, the etch rate non-uniformity, the needed over etch (to remove residual tungsten in case of non planar geometry), and the possible loading effects (see next section), some plugs will see a considerable over etch. Assume that the recession will be 0.2 um in a contact of 0.8 um diameter. The subsequent aluminum metallization will then see an aspect ratio of 0.25. This could possibly lead to aluminum step coverage problems. Different IC manufacturers have different points of views on this subject.

Nevertheless we can illuminate the above stated issue with the following example. Assume an aspect ratio after etch back of 0.25 is allowed and that a film nominal thickness of 500 nm is sufficient to fill the 0.8 μm contacts. Assume further the following "realistic" figures:

absolute deposition non-uniformity:	10%......=50nm
absolute etch rate non-uniformity:	10%......=50nm
over etch:	10%......=50nm
	Total loss: =150 nm

At certain areas of the wafer (where the thinnest film coincides with the fastest etch rate) all these effects will work together giving a total plug over etch of at least 150 nm. Since we also have to include loading effects (see back etch) we come close to what was allowed in the example above (that was 0.2 μm recess). (+/-10% thickness or etch rate spread means a standard deviation of about 3%.)

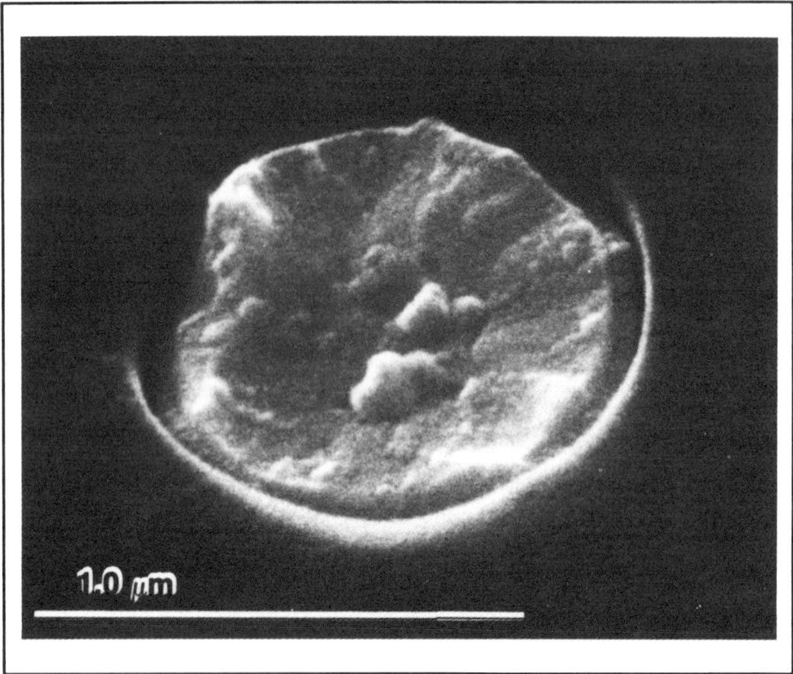

Figure 2.18. Result after etchback using no sacrificial layer. Adhesion layer in this case is TiW. [From Ellwanger et al.[7] reprinted with permission].

2.4 ETCH BACK OF BLANKET TUNGSTEN

At least two distinct approaches can be followed for etching back blanket tungsten; 1) etch back without the use of a sacrificial layer and 2) etch back with the use of a sacrificial layer.

Both methods will be described below.

2.4.1 Etch Back Without a Sacrificial Layer

Despite the fact that a good etch back process is of prime importance for the success of a blanket plug process, not much has been

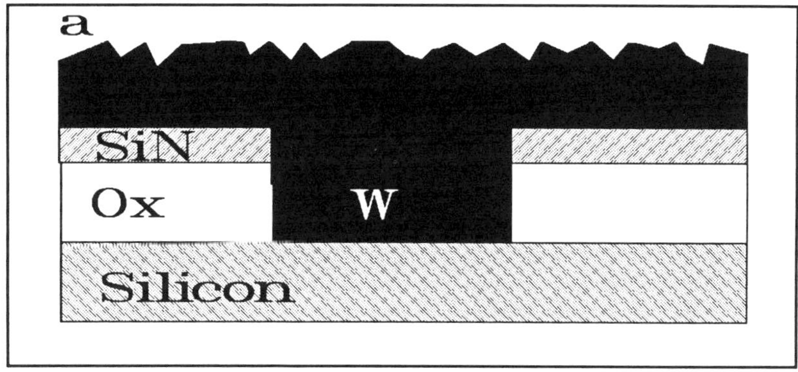

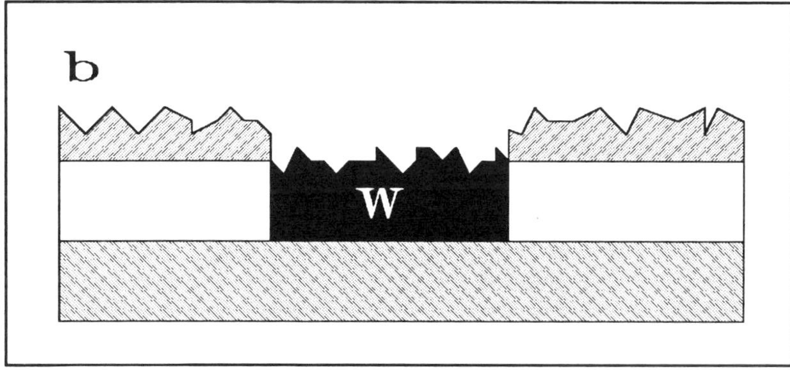

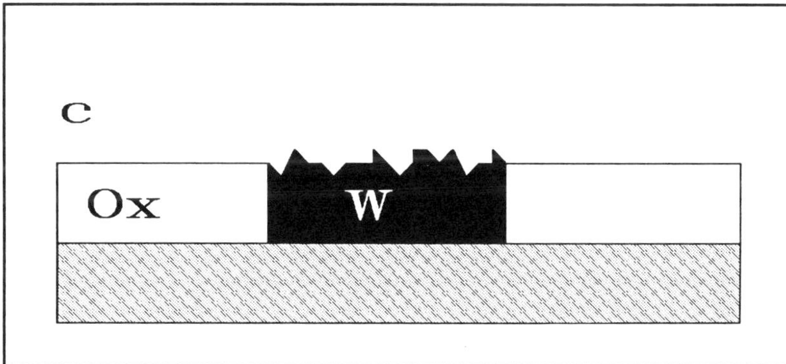

Figure 2.19. Procedure to eliminate loading and oxide surface roughness. After tungsten deposition (a), after tungsten etch (b) and after nitride wet strip (c).

published in this field. Among the reported chemistries are CF_4/O_2, Cl_2/O_2 and SF_6 [Burba et al.[55], Hess[56], Daubenspeck et al.[57], Matsukura et al.[58], Riley et al.[59]]. Only single wafer etch equipment has demonstrated, as of yet, manufacturable processes (see figure 2.18). Some obvious requirements are:

- Good etch uniformity
- Good selectivity towards the dielectric (oxide)
- No etching of the adhesion layer along the contact walls
- Minimum (micro) loading effects
- No etch residues left.

Let us have a somewhat closer look at each of these constraints.

Etch uniformity: This has been highlighted under section 2.3.4. It is clear that for maximal process latitude the etch rate non-uniformity but also the tungsten thickness uniformity should be minimal.

Selectivity: Depending upon what adhesion layer is used different requirements will be needed in terms of etch rate selectivity of tungsten versus the oxide or the adhesion layer. Two cases can be distinguished:

i) The adhesion layer is removed during the etch back and the etch must stop on the oxide (or dielectric material). The tungsten films can exhibit a substantial roughness (ca. 10% of the layer thickness, see also chapter V). In order to prevent that this roughness is imaged into the oxide, (and causing problems in subsequent lithography steps) the etch selectivity of tungsten over oxide should be high. Another important issue is that only a certain loss of the oxide is allowed since a minimum thickness will be required from an electrical point of view. The etch rate selectivity is defined as the ratio of the etch rate of tungsten and the etch rate of the oxide. With optimized (two step) processes it is possible to have sufficient selectivity [Riley et al.[59], Nowicki et al.[60]].

ii) The adhesion layer is not removed during etch back but is used in a subsequent Al deposition step. A good example is TiN. With a proper etch scheme, the selectivity towards TiN can be made high enough thus allow leaving this layer [Körner et al.[259]]. In this case, however, very good uniformities in both the tungsten deposition and the etch back will be necessary.

Etching of the adhesion layer: Generally there will be a different etch rate of the tungsten film and of the adhesion layer. The adhesion layer at the side wall of the contact is susceptible to a faster etch rate due to localized stress. This etching out of the side wall adhesion layer can give micro voids between the tungsten and the oxide wall thus forming a reliability hazard. In severe cases this can even lead to loss of adhesion and consequently loss of the plug.

Loading effects: At the end of the etch back process two mechanisms (so called loading effects [van Laarhoven et al.[61], Berthold et al.[62]]) can dramatically increase the etch rate of the plugs in the contacts:

a) once the bulk tungsten is removed more fluorine is available to etch the plugs and

b) in the case of SF_6 the liberation of oxygen due to the oxide attack also tends to enhance plug etching.

The problem of etched out plugs is that the step coverage of sputtered aluminum will be insufficient if the eventual aspect ratio of the remaining contact becomes too high. The result is an unreliable contact. Two proposed solutions are to change the chemistry (add oxygen) or do a multi step etch process.

Another solution to reduce the loading effect was investigated by van Laarhoven et. al.[61], (see figure 2.19). In their approach there was a 0.3 um PECVD silicon nitride layer deposited atop the oxide prior to the contact opening. The normal procedure of adhesion layer (TiW), tungsten deposition and etch back was followed. Since the nitride etches with about the same rate as the tungsten (selectivity W:SiN=0.8) both the loading is

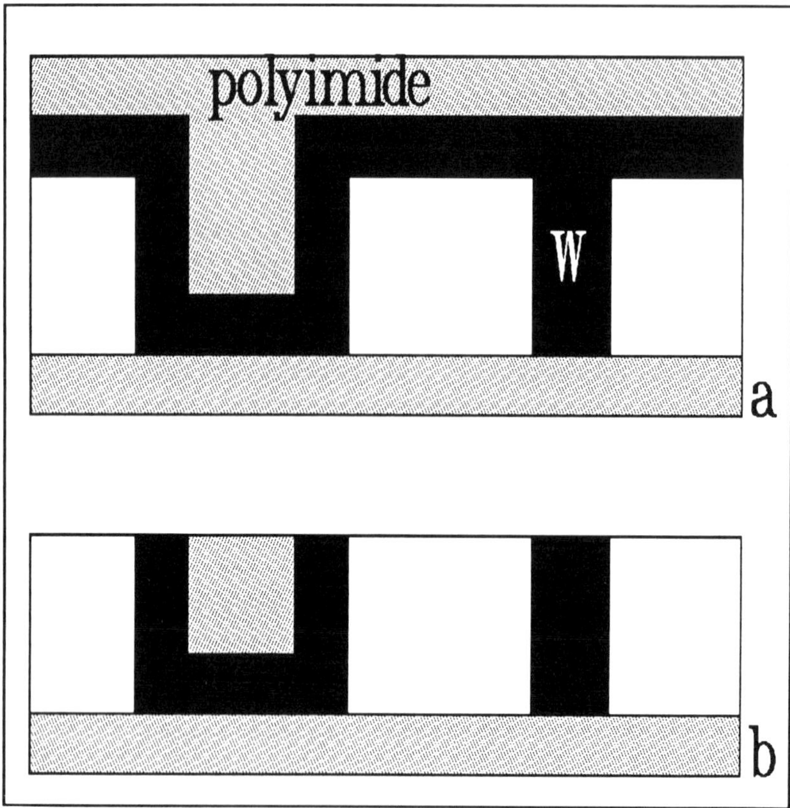

Figure 2.20. Etch back scheme using a sacrificial layer, in this case polyimide. Situation after polyimide spin (a) and after polyimide/tungsten etch back (b).

substantially reduced and moreover the roughness of the tungsten is now translated to the nitride layer. Following etch back, the nitride is removed selectively from the SiO_2, W and TiW in hot phosphoric acid. The results were coplanar plugs and a very smooth oxide surface. The disadvantage of this procedure is that it prevents complete process integration in a cluster tool (see chapter VII).

Etch residues: See section 2.5.

2.4.2 Etch Back Using a Sacrificial Layer

Another way to overcome the problem of tungsten roughness is the use of a planarizing sacrificial layer prior to the blanket etch back step (see figure 2.20) [Higelin et al.[63], Smith et al.[64]]. The trick is to use an etch process which has an etch selectivity of tungsten over the sacrificial material of about one. In this case the roughness of the tungsten is planarized before the etch reaches the oxide level. An additional advantage is that every contact size is allowed. A disadvantage is that the degree of planarization needs to be very high otherwise tungsten strings remain after etch back. Polyimide has been proposed as a candidate for the sacrificial layer. The polyimide can be deposited using a spin on technique similar to resist spinning and is able to give a high degree of planarization. The whole procedure resembles a resist etch back (REB) method for oxide planarization.

2.5 DEGREE OF PLANARIZATION AND THE CONTACT DIAMETER

Before the blanket fill process can be applied, two important parameters need to be considered. These are the variation in the diameter of the contact/via openings and the degree of planarization of the dielectric layers.

The diameter of the contact: Paradoxically there is an upper limit to the contact diameter. This is clearly seen when we look to figure 2.21. Normally one doesn't want to deposit more than ca. 800 to 1000 nm of tungsten (with regard to the cost of WF_6, deposition time, etch time, etc.). This means that the maximum contact diameter is 1 to 1.2 um in order to keep film thickness less than 1 μm or there will be the risk of damage to the large contacts during etch back. This limitation is probably in most cases not that severe since designers can simply replace one large contact by several small contacts. In the case that the tungsten is also used as the interconnect material the problem no longer exists (see chapter V) because the contact

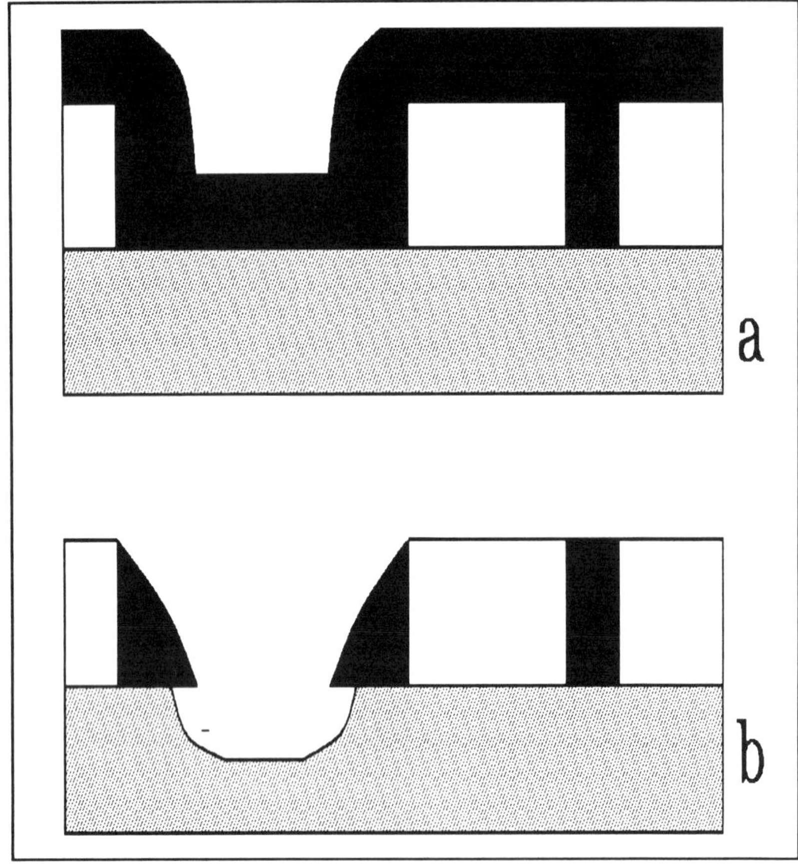

Figure 2.21. Damage of large contact during etch back. In (a) the situation before and in (b) after etch back.

is now protected by the patterned resist.

Degree of planarization (DOP): The blanket tungsten process is in fact a planarization method for contacts and vias. Therefore, if a topography exists before the tungsten deposition is done there will be a risk of tungsten residues or "stringers" after the etch back step (see figure 2.22). This cannot be tolerated since shorts between aluminum lines will occur.

In a detailed study [de Graaf et al.[65]], three different planarization schemes were compared, namely, BPSG flow anneal, Spin On Glass (SOG)

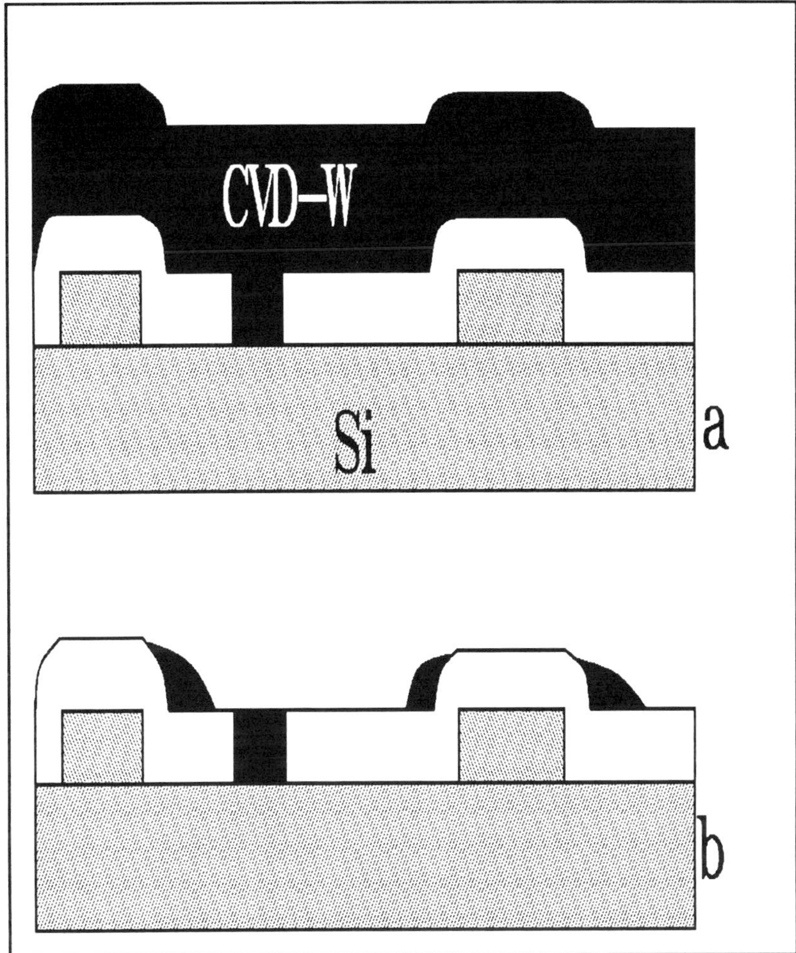

Figure 2.22. Formation of tungsten residues due to insufficient DOP. Situation after tungsten deposition (top) and after etch back (bottom).

and Resist Etch Back (REB). They only studied contact fill and the severest aspect ratio encountered was between two poly lines on top of field oxide (0.9 um wide and 0.8 um deep). It was found that if the maximum step height after planarization was 0.2 um no tungsten residues remained. Only SOG and REB planarizations could meet this requirement. This result will vary of course with each specific topography and the typical etch back process used but it does illustrate that for the blanket tungsten contact fill a good degree of planarization is necessary(75% in this case). This planarization is not only needed for the blanket tungsten step but is also required because of focusing problems in post lithographic steps and step coverage problems in subsequent dielectric films. It is therefore not valid to make this requirement for planarization specific to the blanket tungsten process only.

2.6 BLANKET TUNGSTEN MATERIAL CHARACTERIZATION

The following tungsten properties are important with respect to IC implementation:

- Stress
- Bulk resistivity
- Density
- Roughness (grain size)
- Purity

Since many of these parameters are of more importance when tungsten is applied as an interconnect, we refer you to chapter V for further discussion.

CHAPTER III

THE SELECTIVE TUNGSTEN APPROACH

The attractive features of a selective tungsten process in IC's were mentioned more than 10 years ago. Two problems have prevented a timely implementation:

i) the attack of the silicon substrate by WF_6 and

ii) the lack of understanding and control of the loss of selectivity.

For both issues substantial improvements have been made:

i) the introduction of silane in the gas phase can effectively suppress the silicon attack and

ii) the use of cold wall reactors and appropriate pretreatment techniques have significantly improved selectivity.

In this chapter we will discuss what chemistries have been tried, what problems have been found and what solutions have been proposed. Also such issues as barrier stability, contact resistance and leakage current will be mentioned.

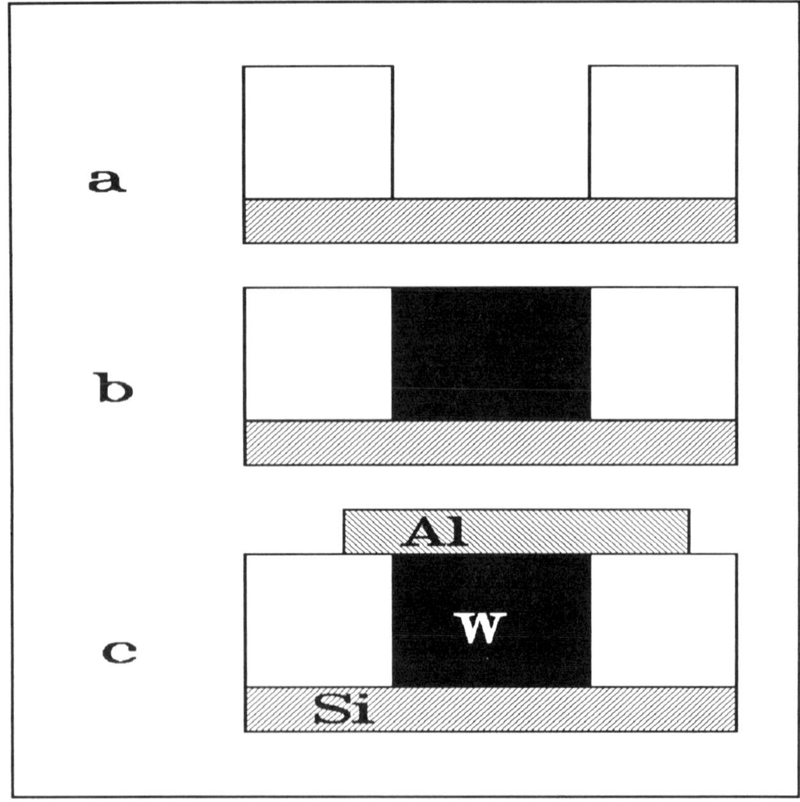

Figure 3.1. After opening of the contacts or the vias (a) tungsten is deposited selectively (b) followed by the sputter deposition of aluminum (c).

3.1 PRINCIPAL STEPS

The selective process relies on the fact that there can be a substantial difference in the nucleation rate on different substrates. Tungsten tends to nucleate much easier on (semi) conductors such as silicon, silicides, aluminum and TiW than on dielectric materials. Implicitly we say then that the selectivity is not infinite: after some time the selectivity is "lost" (see section 3.5). However, in many cases it is possible to obtain a fair amount of selectivity and layers as thick as 3 um without a significant loss of selectivity have been reported [Wilson et al.[66]].

The selective tungsten process to fill contacts or vias is in principle much simpler than the blanket approach: No adhesion layer and etch back is needed in order to arrive at planar plugs (however, there are also some limitations, see chapter IV). In figure 3.1 we sketched the different key steps in the process. The problem to develop a "generic" selective tungsten process is that there can be a large variety in the "contact" material (silicon, silicides, aluminum and other materials) but also in the dielectric materials (thermal oxide, plasma oxides, TEOS based oxides, Spin On Glass (SOG), silicon nitride, polyimide). In addition the oxides can be doped with phosphorus and/or boron. With regard to all these possibilities it might be a good idea to first briefly review some properties of these materials and their compatibility with the selective tungsten chemistry.

3.2 TYPES OF SUBSTRATES

Obviously there is a distinct difference between the contacts and the vias. In the case of the contacts we can encounter:

- Mono crystalline-silicon. Of course this can be n or p-type. Several problems have been reported such as encroachment (see below) and differences in (apparent) growth rate on n versus p-type.

- Poly crystalline-silicon. Mostly n-type, however, in contemporary BICMOS and CMOS processes p-type can be present as well.

- Silicides. These materials are applied because of their low resistivity. Because of this property and compatibility with post process steps (oxidations [Murarka[86]], oxide depositions, or doping procedures [van den Hove et al.[87]]) silicides are widely used to lower the sheet resistance of shallow junctions and poly gates. The most commonly used for polycide applications are $MoSi_2$ and WSi_2 (see chapter IX), but other possibilities (salicide applications) are $TiSi_2$, PtSi, and $CoSi_2$ [Verhaar et al.[88]].

- TiN or TiW. TiN can be easily formed in a salicide step during the nitridation of either Ti or TiSi$_2$ [Tang et al.[90]]. TiW [Wolters et al.[89]] and TiN [Tang et al.[90]] have been reported for use in strap applications (local interconnects).

As the integration goes on it can be envisioned that the role of the mono-silicon/tungsten interface becomes less since the device performance demands more and more the technique of cladding the shallow junction areas with silicides.

In the case of the vias we deal with the following materials:

- Aluminum. This is by far the most commonly used interconnect material. It can be doped with elements such as Si [Learn[93], Hirashita et al.[92]] and Cu [D'Heurle[91]] to improve the properties such as contact reliability and electromigration. The maximum wafer temperature allowed once aluminum is present is about 400-430^{0}C.

- Tungsten. In ULSI type of circuits there is a trend to more often use tungsten as the interconnect material. See also chapter V.

- Various other materials such as gold [Haberle et al.[96]], molybdenum [Oikawa et al.[94]] or copper [Ting et al.[95]] which can also be used as the interconnect material.

- Sometimes there is a need for anti reflective coatings (ARC) atop the aluminum to prevent lithographic problems. Materials reported are amorphous silicon and sputtered TiW or TiN.

In the near future we will see that the use of aluminum will diminish and other interconnect materials will come into the picture. Nevertheless most studies of selective tungsten in vias are presently focused on aluminum.

3.3 TYPES OF DIELECTRIC LAYERS

As mentioned briefly above there are many dielectrics in use:

- BPSG based on TEOS decomposition or SiH_4 oxidation. The phosphorous and boron concentrations can show quite a range (P: 0-8 wt%, B: 0-8 wt%).

- Plasma oxide based on SiH_4 or TEOS and doped or undoped. Mostly used atop aluminum with regard to the limited temperature budget.

- SOG. Here too a wide range of materials is available. Because of its good planarization characteristics it is becoming more popular [Yen et al.[97], Chen et al.[98]].

- Silicon nitride by both thermal or plasma activation.

- Organic materials. A pertinent example is spin-on polyimide. Although having been in use for a long time this material has recently received more attention [Delfino et al.[99], Pattee et al.[100]].

It is well established that the degree of selectivity can vary extremely going from one dielectric to the other [Chow et al.[67], Broadbent et al.[131], Chow et al.[133], Bradbury et al.[134], Wilson et al.[135]]. Silicon nitride, for instance, is notorious for its low selectivity whereas phosphorous doped glasses show an improved selectivity compared with undoped glasses (see also section 3.5).

3.4 CHEMISTRY OF SELECTIVE TUNGSTEN

Before we go more deeply into the chemistries currently in use to deposit selective tungsten we have to better define the concept of selectivity. Selectivity loss is probably best defined by the number of tungsten crystallites and the size distribution present on the dielectric after deposition. Unfortunately this is not easy to quantify in practice. Standard particle counters cannot be used because they can not discriminate between tungsten due to loss of selectivity and tungsten present in the contacts/vias. Although new developments using low angle incident beam techniques may change this, this concept needs further evaluation. Therefore, for an in line selectivity check, one needs to do the painful exercise counting the tungsten nuclei using an instrument such as a dark field microscope. The disadvantage to this method is that only a limited surface area can be inspected and categorization into size is not possible . In addition the translation of those counts into killing power (for yield) is not a trivial matter. Nevertheless, relative comparisons are attainable. It is also important to note that the selectivity depends of course on the deposition time or the thickness of the tungsten. Therefore, a certain selectivity number should always be correlated to a tungsten thickness.

In an effort to quantify the dark field counts, Chow et al.[67] compared these counts with two other techniques, RBS and Total Reflectance X-Ray Fluorescence (TRXRF). Some discrepancies between these techniques were found. See section 3.5 for more details. To the authors knowledge no relationship between yield and actual selectivity loss has yet been reported.

Several chemistries are possible to deposit selective tungsten. Tungsten sources like WF_6, $W(CO)_6$ and WCl_6 have been reported to give tungsten films [see chapter VI]. H_2, SiH_4, B_2H_6, Si_2H_6, PH_3 [Ohba et al.[68,69]], SiH_2Cl_2 [Herd et al.[70]] and other gasses have been evaluated to reduce the tungsten source. Today only two chemistries are seriously studied with respect to selective tungsten deposition: H_2/WF_6 and SiH_4/WF_6. Of these the hydrogen based chemistry was the first reported to give selective tungsten deposition and we will start our discussion with this chemistry.

3.4.1 H_2/WF_6 Chemistry

The displacement reaction: Although it is believed that the key reason for selectivity is the activated adsorption of hydrogen on tungsten there is obviously no tungsten available in the beginning of the deposition (in a contact to silicon). The first reaction which will occur is that between silicon and WF_6, the so called displacement reaction [Broadbent et al.[44]]:

$$2WF_6 \; + \; 3Si \; \text{------>} \; 2W \; + \; 3SiF_4 \qquad (3.1)$$

This reaction is selective since almost no reaction with SiO_2 will take place (assume oxide as the dielectric). The reaction can only take place as long as Si is available for the reaction between WF_6 and the silicon. This reaction was shown to be very fast, within 6 seconds tungsten growth stops. After a certain thickness of tungsten is formed, the tungsten film starts to act as a Si diffusion barrier: the reaction is self limiting. The thickness of this self limiting layer (see Hitchman et al.[75] for a review) has been the subject of several studies and has been reported to range from as thin as 10 nm [Ahn et al.[71]] to almost 1.5 micron [Kobayashi et al.[186]]. In an elegant study of surface reactions, Yu et al.[75] found that reaction 3.1 is only valid at temperatures lower than about 400°C (typical for selective depositions). Above that temperature the main reaction products are tungsten and SiF_2:

$$WF_6 \ + \ 3Si \ \text{------>} \ W \ + \ 3SiF_2 \ (T>400°C) \qquad (3.2.)$$

This was confirmed in another study using RBS spectroscopy by Kuiper et al.[72]. We see that in this case the silicon consumption is twice that of reaction 3.1. The presence of H_2 has only a negligible effect on the course of the reaction since the reaction between Si and WF_6 is so much faster than that between H_2 and WF_6 [Broadbent et al.[44]]. In the remaining part of this section we will summarize some explanations offered in the literature for the observed thickness range.

There is evidence that the reaction between the silicon and WF_6 proceeds at the tungsten surface and that the diffusion of silicon through the tungsten is necessary to sustain the reaction. Since tungsten is an effective barrier against Si diffusion (at temperatures below 600°C [Pauleau et al.[109], Thomas et al.[110]]) films thicker than about 10nm cannot easily be explained with a silicon diffusion mechanism. It has been proposed and experimentally confirmed that the as-deposited tungsten is rather porous (about 75% of the normal density) [Kuiper et al.[72], Kobayashi et al.[186]]. The channel formation will be enhanced because each tungsten atom will replace between 1.5 to 3 atoms of silicon [Kuiper et al.[72]]. This porous structure could be allowing silicon to diffuse through the micro channels. After a certain time, however, the channels will become plugged by tungsten.

Another explanation is that the presence of the native oxide (in which there are pinholes) can cause thicker films of tungsten [Green et

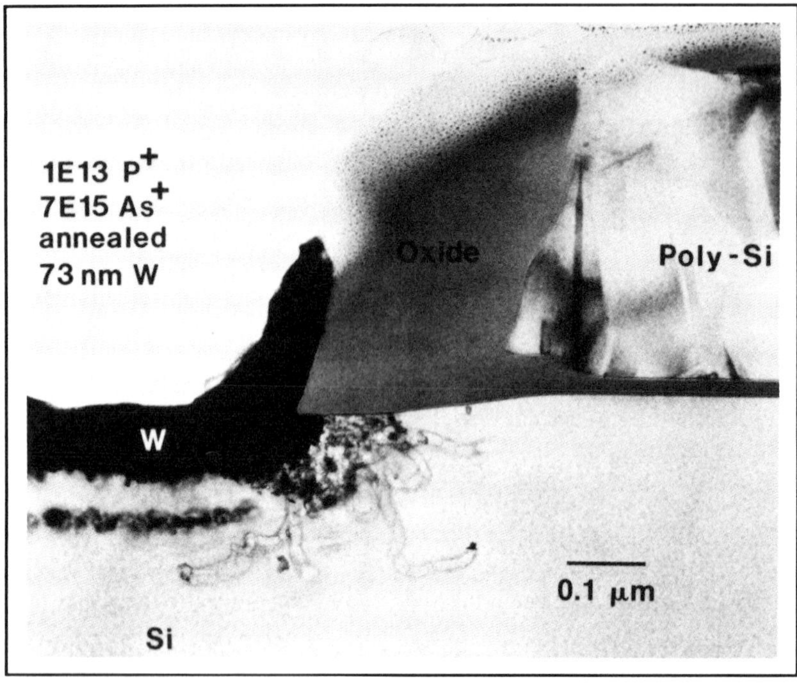

Figure 3.2. Severe wormhole formation in the source area of a MOS transistor due to selective tungsten deposition. [TEM courtesy J. Flanner, Signetics Corporation].

al.[73]]. The WF_6 penetrates first through the pinholes and replaces the silicon with tungsten (see figure 3.3). Thus it is expected and experimentally found [Hitchman et al.[74] and references therein] that the surface preparation has an effect on the self-limiting tungsten thickness. The better the quality of the (native) oxide (less pinholes) the thinner and smoother the limiting film.

In a detailed analysis Hitchman et al.[74], came to an interesting conclusion and pointed out that trace amounts of water can influence the result tremendously. For instance, WF_6 will not react with SiO_2 in a dry ambient possibly due to the formation of a protecting WO_3 film on the oxide. However, if water is present there will be a reaction between WF_6 and H_2O according to:

$$WF_6 \; + \; H_2O \; \text{------->} \; WOF_4 \; + \; 2HF \qquad (3.3)$$

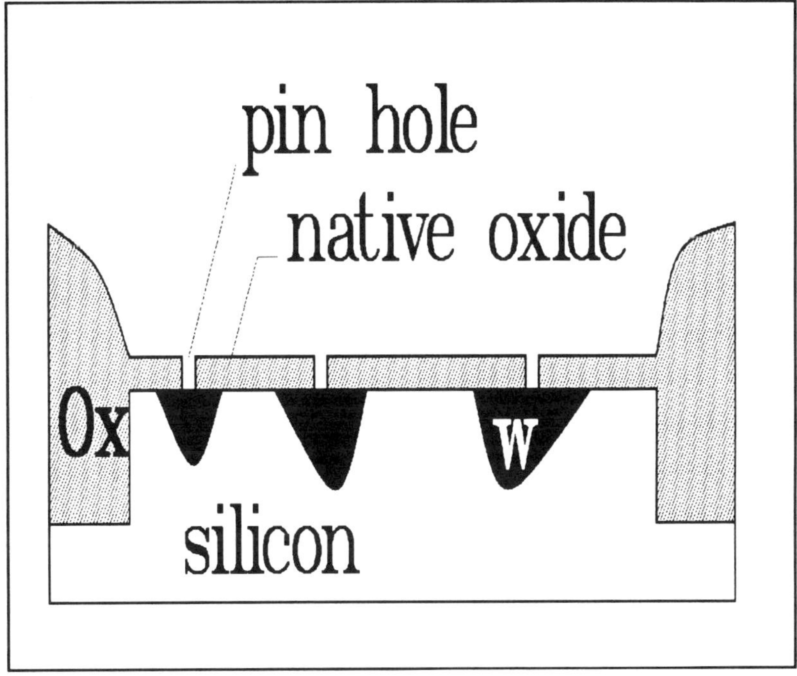

Figure 3.3 Penetration of WF$_6$ through pin holes in the native oxide. This causes thicker tungsten films than in the case that no native oxide would be present.

HF can then react with WO$_3$ and the oxide surface becomes available for further attack. Therefore, different moisture levels in reactors can lead to different results in the tungsten thickness for the encroachment reaction.

Kuiper et al.[72], suggested a relationship between the thickness of the limiting layer and the type of reactor used: cold wall or hot wall. The key here is again that the water content in non-loadlocked hot wall reactors will be much higher than in a cold wall reactor.

Although the reaction between silicon and WF$_6$ is necessary to start the tungsten deposition, at the same time it causes many problems and in fact has prevented the successful implementation of selective tungsten based on the H$_2$/WF$_6$ chemistry. TEM studies of the tungsten-silicon interface

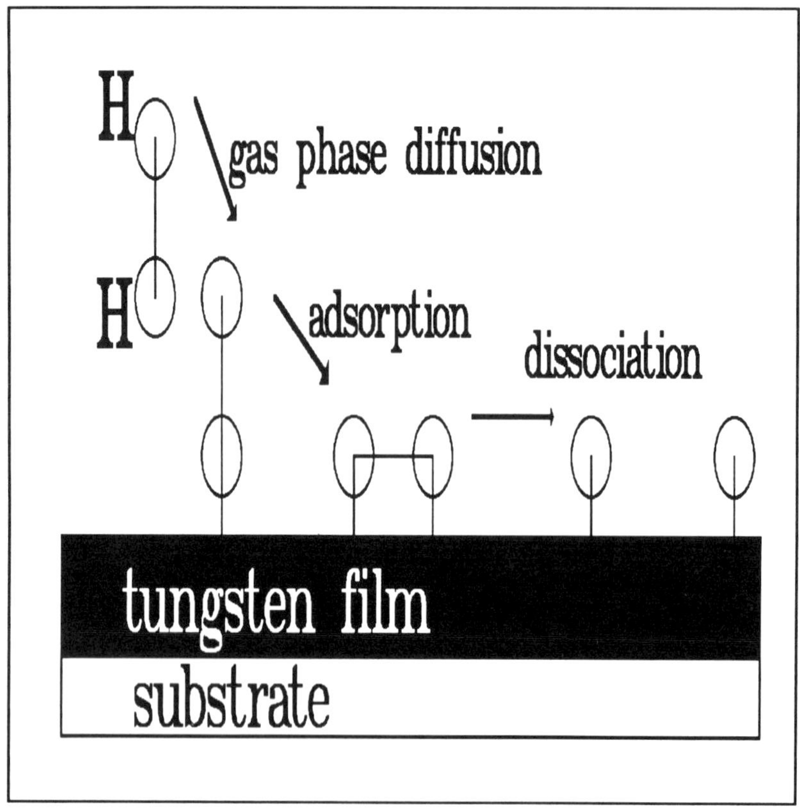

Figure 3.4. Adsorption and activation of hydrogen on the tungsten surface.

show very rough interfaces. In addition, long (> 1um) tunnels ("wormholes" see figure 3.2) in the silicon can be observed [Stacy et al.[11], Levy et al.[76], DeBlasi et al.[77], Blewer et al.[77]]. Especially shallow junctions are sensitive since above phenomena can lead to high junction leakage currents. The encroachment can also consume the heavily doped silicon material atop the junction such that the contact resistance is no longer acceptable and shows large fluctuations ($R_c \sim \exp[1/\sqrt{N_D}]$, where N_D is the surface dopant concentration). It has been shown by Levy et al.[76], that the encroachment can be suppressed (albeit at the cost of growth rate) by the addition of SiF_4 since this will force reaction 3.1 to shift to the left. Furthermore, recently it has been found that the introduction of SiH_4 in the gas phase can also reduce the silicon consumption completely (see section 3.4.2).

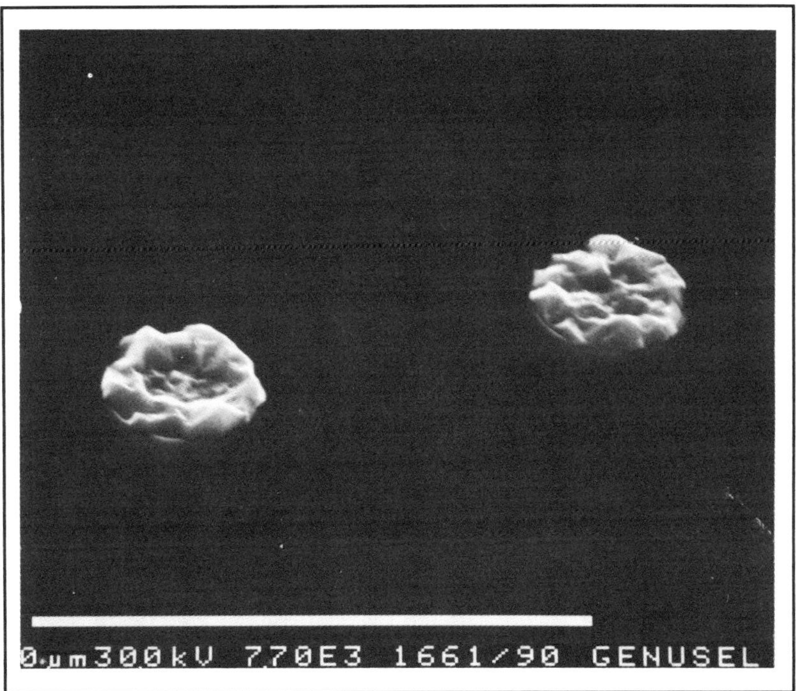

Figure 3.5. Top view of contacts filled with selective tungsten (H_2/WF_6 chemistry).

The selective H_2/WF_6 step: Once tungsten is formed as discussed above hydrogen will adsorb at the tungsten surface. The adsorption of hydrogen on metal surfaces is well studied and it is known that the (stable) H_2 molecule is activated because it dissociates into adsorbed atomic hydrogen and becomes in this way prone to further reaction with WF_6 (see figure 3.4). In order to allow the reaction to proceed the thermodynamically stable WF_6 has to adsorb also and in subsequent steps, 6 fluorine atoms will be removed. The reaction has been kinetically characterized in both hot wall [Broadbent et al.[44], Pauleau et al.[45]] and cold wall [McConica et al.[46]] systems with essentially similar results. The deposition rate was found to follow the expression:

$$\text{Rate} = A \times e^{-Ea/RT} \times P_{WF6}{}^{0} \times P_{H2}{}^{1/2} \tag{3.4}$$

where A is the pre-exponential factor. The activation energy appears to be about 0.7 eV.

The knowledge of the reaction orders and the activation energy allows in principle the determination of the rate determining step (RDS). Originally it was proposed that the dissociation of hydrogen at the tungsten surface was the RDS:

$$H_{2,a} \text{-------> } 2H_a \tag{3.5}$$

where the subscript "a" identifies the adsorbed species. McConica et al.[46], showed, however, that another possibility exist, namely, the desorption of HF from the tungsten surface. In this case one arrives at a slightly different rate equation:

$$\text{Rate} = A \text{ x } e^{-Ea/RT} \text{ x } P_{WF6}^{1/6} \text{ x } P_{H2}^{1/2} \tag{3.6}$$

A small order like 1/6 is not easy to detect experimentally. Therefore equation 3.6 can still be in agreement with the observed "zero" order of WF_6.

Since the activation of hydrogen does not occur on dielectric surfaces like SiO_2 or Si_3N_4 the reaction will proceed only where metal is already available. This is the explanation for selectivity. In section 3.5 we will discuss reasons for selectivity loss.

In the case where a silicide is the substrate there is not much known about the initial step: a reaction of WF_6 with the silicide under formation of W or an adsorption of hydrogen to form activated atomic hydrogen. It can be shown [Hårsta et al.[78]] that thermodynamically the following reactions can proceed:

$$11WF_6 + 6TiSi_2 \text{------> } 11W + 6TiF_3 + 12SiF_4 \tag{3.7}$$

and

$$3CoSi_2 + 2WF_6 \text{------> } 2W + 3CoSi + 3SiF_4 \tag{3.8}$$

Equation 3.7 only holds for temperatures lower than 500°C. Above that temperature TiF_3 is no longer predicted and experimentally verified [Smith et al.[114], Ng et al.[115]].

Selective deposition also can be obtained on TiN. The reaction between TiN and WF_6 is thermodynamically favorable:

$$2TiN \ + \ WF_6 \ \text{------>} \ W \ + \ 2TiF_3 \ + \ N_2 \qquad (3.9)$$

The deposition, however, can exhibit severe initiation times (10 min.) [Rana et al.[8]]

On Al and TiW no surface reaction is needed since hydrogen can adsorb directly and become activated by dissociation. In the case of aluminum the preclean step is critical. Uncleaned aluminum gave no deposition while a dip in HCl showed good results [Ng et al.[115]]. See the review article from Broadbent[263] for more details.

3.4.2 SiH_4/WF_6 Chemistry

Until recently the silane based chemistry was only in use for blanket W depositions [Fuhs et al.[80]] (see chapter II). These blanket depositions were done at temperatures between 400 and 500°C. During the 1988 Workshop on Tungsten and Other Refractory Metals in Yorktown Heights however, several investigators announced that at temperatures of about 300°C selective tungsten depositions can be obtained using the silane chemistry [Foster et al.[81], Kusumoto et al.[82]]. In addition no encroachment or tunnel formation was found when filling contacts with tungsten (however, there is probably still some silicon loss, see Itoh et al.[85]). In another study [Ellwanger et al.[113]] it was found that the interfacial fluorine content for CVD-W/Ti and CVD-W/$TiSi_2$ was orders of magnitude lower for silane based chemistry than for hydrogen based chemistry. As an additional advantage very high growth rates such as 500 nm/min were reported. This last feature opened the way for a single wafer reactor approach.

The chemistry of the SiH_4/WF_6 system is, however, rather complicated. It seems appropriate at this point to illuminate some of the particular properties of this chemistry.

Table 3.1

Comparison between observed and predicted phases

SiH_4/WF_6	X-ray	Thermodyn. pred.
0.55	α-W	α-W
1.1	α-W	α-W
1.3	α-W	α-W
1.5	β-W+amorph.	α-W
1.6	β-W+amorph.	α-W
2.0	amorph.+α-W	W_5Si_3+α-W
2.5	amorphous	W_5Si_3+WSi_2
3.0	amorphous	W_5Si_3+WSi_2
3.8	amorphous	WSi_2+(W_5Si_3)
15	no reflections	WSi_2+Si

Deposition temperature = 270 ^{o}C; Total pressure = 200 mTorr,
data from reference 83, reprinted with permission.

Film composition and texture as a function of SiH_4/WF_6 flow ratio: In this
section we discuss the film composition and texture as a function of the
SiH_4/WF_6 chemistry. In a study by Schmitz et al.[83], the SiH_4/WF_6 flow ratio
was varied between 0.5 and 15 while keeping the SiH_4 flow constant. A first
impression can be gathered from the X-ray diffraction spectra (see figure
3.6). Only for flow ratios smaller than 1.3 is a stable low resistivity α-W
phase obtained. Between 1.5 and about 3 we see a diffraction pattern which
can be identified with that of the meta-stable high resistivity β-W phase
[Tang et al.[117]]. In addition, a broad peak appears with a high at the 210
reflection of β-W (which coincides with the 110 reflection of α-W),
characteristic of amorphous or micro-crystalline material. At higher flow
ratios all diffractions disappear and at a ratio of 15 no peak is observed.

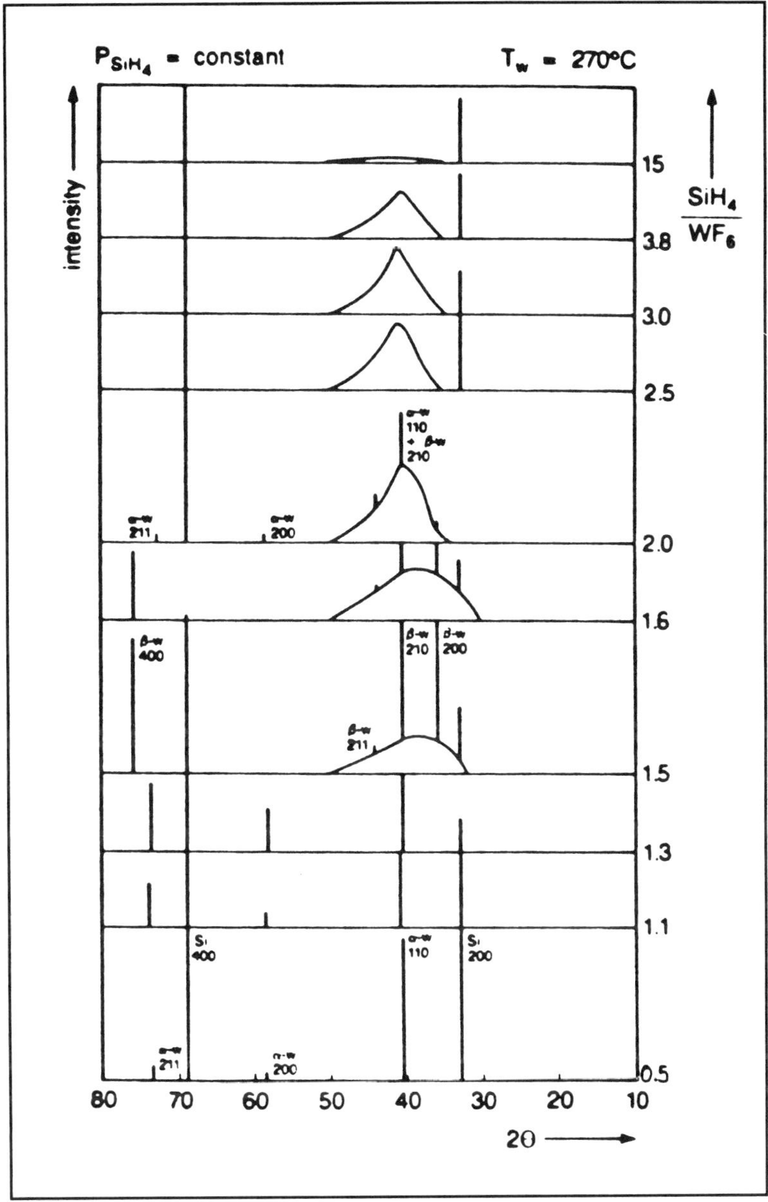

Figure 3.6. X-ray diffraction spectra of films deposited at several SiH$_4$/WF$_6$ ratios. Temperature 270°C, pressure = 200 mTorr. [From ref 83, reprinted with permission].

In table 3.1 the data of the X-ray spectra is gathered and compared with solid state phases as predicted by thermodynamic calculations. There appears to be a reasonable agreement between the thermodynamic predictions and the experimental (X-ray) data. Of course we should keep in mind that because of the low deposition temperature, the silicide phases stay micro-crystalline. Therefore, we see no silicide reflections. In disagreement with the prediction is the formation of the meta-stable β-W phase at the 1.5-1.6 flow ratio. However, after annealing the β-W sample at 800°C in vacuum, α-W reflections plus those of W_5Si_3 appear. Thus the reaction proceeds far from equilibrium and the beta-W phase is only kinetically stable. The formation of β-W and its conversion into α-W is described extensively in the literature [Morcom et al.[116], Tang et al.[117], Paine et al.[118], Davazoglou et al.[119]].

Table 3.2

Calculated and experimental composition and tungsten resistivity

flow ratio SiH_4/WF_6	<---Auger at%--> W	Si	O	F	Auger Si/W	calc. Si/W	res. uΩcm
0.55	94	3	3	<1	0.03	0	19
1.1	95	3	2	<1	0.03	0	-
1.3	93	3	2	<1	0.03	0	100
1.5	87	7	6	<1	0.08	0	430
1.6	82	9	9	<1	0.11	0	390
2.0	63	35	<1	<1	0.56	0.32	-
2.5	53	45	<1	<1	0.85	0.66	201
3.0	51	47	<1	<1	0.92	1.18	230
3.8	49	49	<1	<1	1.00	1.96	110
15	43	55	<1	<1	1.28	12.8	250

At 270°C and 200 mTorr. Data from reference 83, reprinted with permission.

More insight can be obtained from the elemental film composition as seen in table 3.2. Several interesting conclusions can be drawn on comparing the actual film composition with the calculated composition (from thermodynamic data). In the flow range 0.55-1.3 we see reasonable agreement as almost pure tungsten is obtained. In the range 1.5-2.5 we see that the calculation <u>underestimates</u> the actual Si concentration. This is in the flow ratio range where we see the broad amorphous (probably β-W) peak. It is known that the β-W lattice is stabilized by impurities like fluorine [Tang et al.[117]] or oxygen [Busta et al.[120], Hagg et al.[121]]. In this case the impurity is mainly silicon. Nevertheless we see a gettering of oxygen (acting as another stabilizing impurity) at ratios 1.5 and 1.6 where we also observe the distinct β-W reflections. In the range of 3.0-15 we see that the calculation <u>overestimates</u> the Si content. This is because starting at a ratio 3.0 there is more SiH_4 present than can be consumed by the WF_6. Since SiH_4 is an unstable compound thermodynamics predicts that it will decompose totally into Si and H_2 at equilibrium. At 270°C, however, this reaction is kinetically blocked and therefore explains the silicon content overestimation.

The resistivity of the α-W phase in this study is more than twice the values obtained at higher deposition temperatures (>400°C) [Fuhs et al.[80]]. This is probably due to the relatively high silicon content (however, see section 5.3.5). The increased resistivity for selective tungsten is in most applications not a problem.

Selectivity: The most important parameters for selectivity for the SiH_4/WF_6 chemistry are the temperature and the reactant flow ratio. Although there is some dispute on how to determine exactly the wafer temperature (see section 7.3), there is a general belief that the selective temperature window is rather narrow (270-320°C). Below about 250°C there is no growth at all and above 350°C the selectivity is completely lost, as only blanket depositions are observed. See section 3.5 for more details about loss of selectivity.

Kinetics: An important part of a kinetic study is normally to characterize the growth rate as a function of partial pressures of the reactants and/or

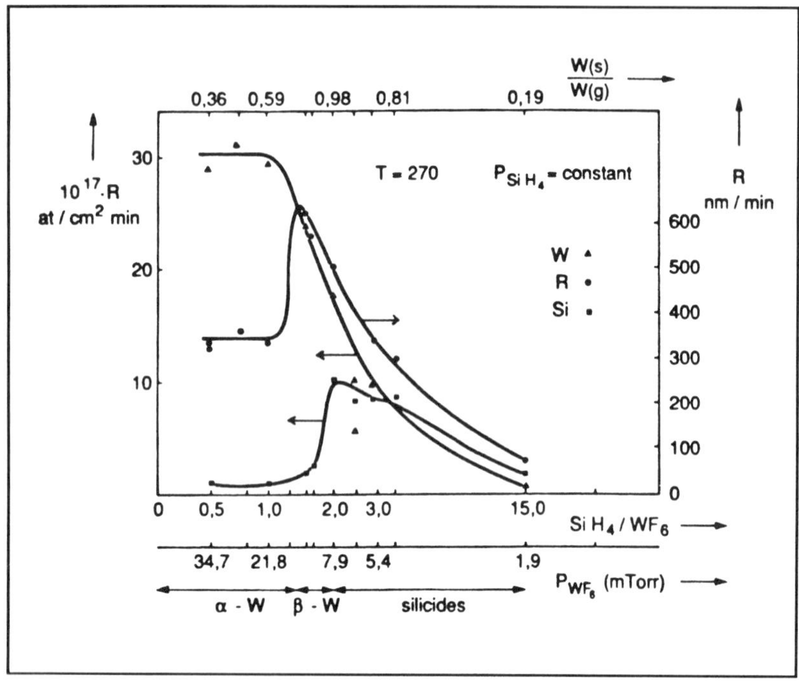

Figure 3.7. Film growth (R in nm/min) and W and Si dep. rate (W and Si in at/cm^{2}sec). W(s)/W(g) is the conversion degree of WF$_6$. [From ref. 83, reprinted with permission].

products, and as a function of the temperature. An implicit assumption made is that in the studied pressure and temperature range the composition of the film does not change. Unfortunately this is not valid in the range of reactant ratios we study, as is obvious from table 3.2. In the following discussion it will be shown that more insight can be obtained by studying the Si and W growth rate separately [Schmitz et al.[87]].

Let us have a look to the graph in figure 3.7. The total growth rate R (determined from SEM cross sections in nm/min) and the W and Si growth rates (determined from RBS spectroscopy in at/cm^2 min) are given as a function of the WF$_6$ partial pressure. For convenience the reactant ratio is also given. Across the upper axis the conversion of WF$_6$ (W(s)/W(g)) is given. For instance, a value of 0.36 means that 36% of the incoming WF$_6$ is consumed in the reaction. We see that very high conversion numbers are obtained. It is good practice to keep the conversion number at least lower than 10% in kinetic studies. In that case one can calculate the reactant partial pressures from the reactant flows (sccm) and the total pressure thereby assuming a gradientless reactor. Also, the influence of any reverse reaction generated by the product formed can be neglected. However, in the SiH_4/WF_6 chemistry very high growth rates can be observed (up 1000 nm/min [Foster et al.[81], Kusumoto et al.[82]]) making it difficult to study this chemistry in standard CVD equipment. Very large flows (and thus large pumps) are required to keep the conversion factors low. Nevertheless interesting conclusions can be drawn from figure 3.7.

On going from 0.5 to ca. 1.3 (ratio) we see that the overall growth rate is constant as is the case with the tungsten growth rate. Since the Si content is very low the Si growth rate is almost zero. Kinetic studies done in this reactant ratio (keeping the conversion factor low) indeed show the following rate law [Schmitz et al.[48], Rosler et al.[49]]:

$$\text{Deposition rate} = \text{constant} \times [P_{WF6}]^0 \times [P_{SiH4}]^1 \tag{3.10}$$

(see also figure 3.9a,b). The reaction mechanism is probably much more complicated than equation (3.10) suggests. An Arrhenius plot taken at a SiH_4/WF_6 ratio of 0.75 shows that an unambiguous activation energy cannot be obtained (see figure 3.8).

If the ratio is further increased, there is a maximum in both the overall film growth rate (R) and the Si rate but not in the tungsten deposition rate. This is the range where the X-rays shows β-W reflections. Since the relatively low Si content cannot account for the huge increase in growth rate and because the tungsten growth rate even decreases, this implies that the maximum in the overall growth rate can only be explained by a change in density. Indeed a determination of the density of the film deposited at a reactant ratio of 1.6 shows a density of 10 gr/cm^3 (which is close to what has been reported for β-W [Morcom et al.[116]]). For the α-W phase a density of ca. 19 gr/cm^3 was found (bulk tungsten is 19.3 gr/cm^3).

On increasing the ratio further (implying lower WF_6 flow) we enter the silicide regime and both the Si and W growth rates drop. See reference 83 for more details.

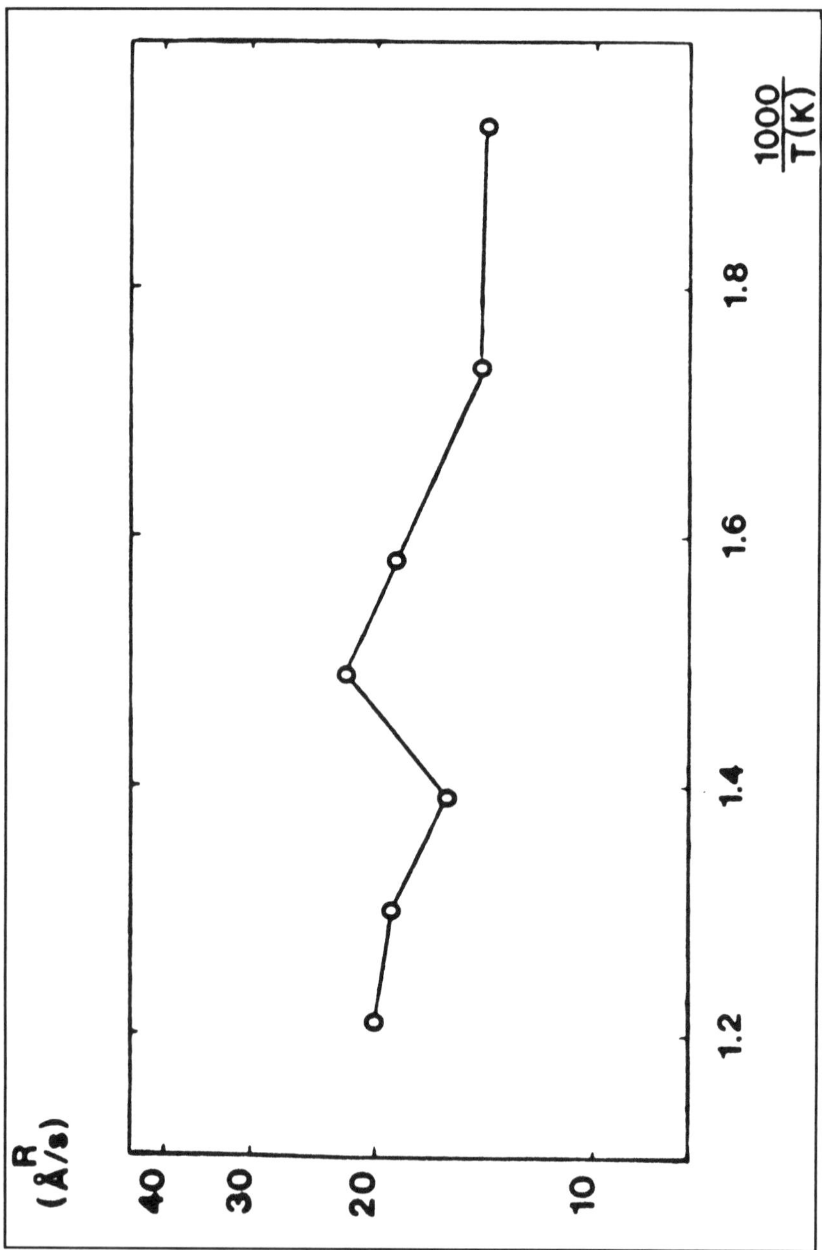

Figure 3.8. Deposition rate as a function of the temperature [From ref. 48, reprinted by permission of the publisher, The Electrochemical Society, Inc.].

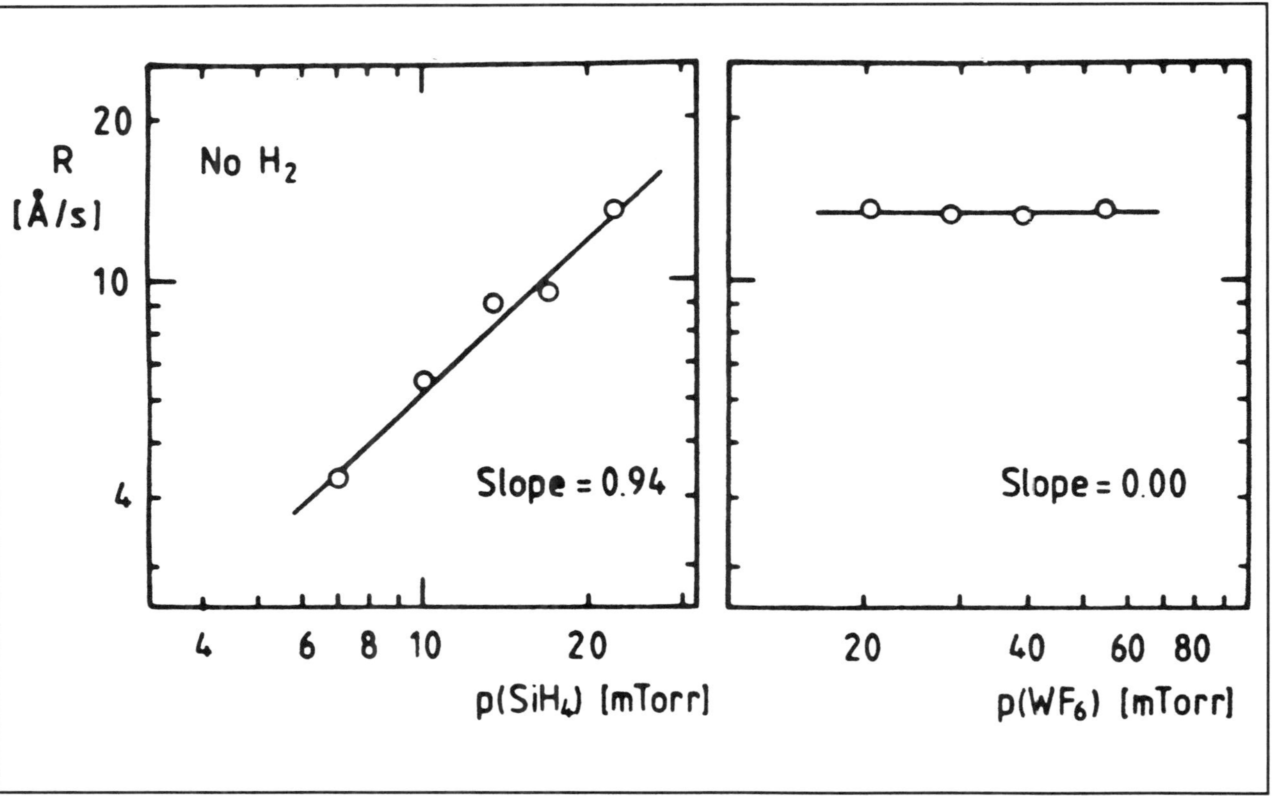

Figure 3.9.a. Order of the reactants for the SiH_4/WF_6 chemistry at 360°C.

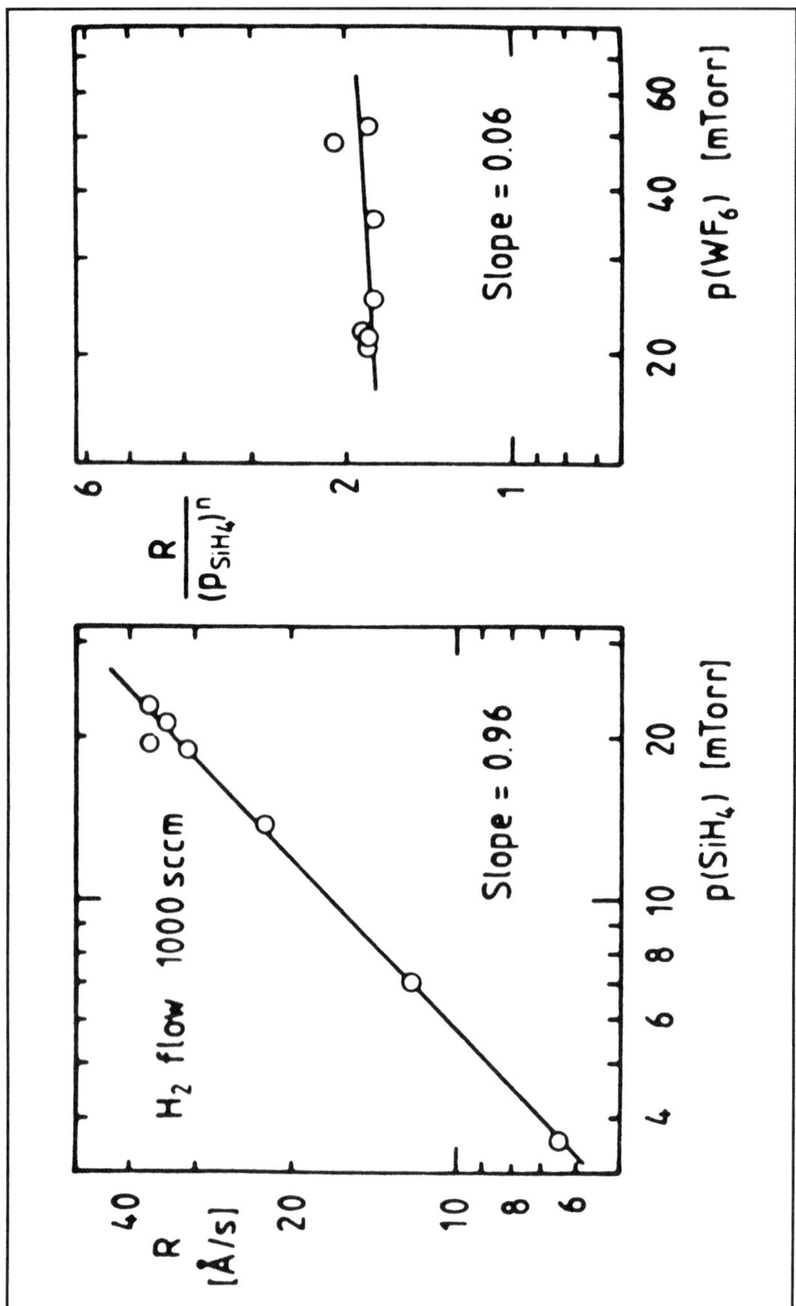

Figure 3.9.b. Same as figure 3.9.a. but now for the $SiH_4/H_2/WF_6$ chemistry at 430°C. [From ref 87, reprinted by permission of the publisher, The Electrochemical Society, Inc.].

3.5 MECHANISMS OF SELECTIVITY LOSS

Several reasons are known to be responsible for selectivity loss. Although much effort has been put forth to unravel the mechanism, we are still far from a complete understanding of the phenomenon. In the following we will discuss some facts about selectivity loss.

Selectivity in terms of reaction rates: When we say that a certain reaction or deposition is selective we mean that the reaction rate for unwanted side reactions (here nucleation on oxide) is slow compared with the wanted reaction route ie. the nucleation on silicon or metals. Typically after a long enough (nucleation) time tungsten growth will begin on the dielectric layer. The crux is, of course, to develop a procedure (i.e. a pretreatment + selective deposition + a post treatment) which can give thick enough tungsten before the loss of selectivity gives killing defects. But, even in the event of selectivity loss this is not necessarily killing: clearly grains of 0.1 um will cause less yield loss as 1.0um grains can.

Surface contamination: Surface contamination can be a major reason for premature loss of selectivity. Such contamination can be, for instance, particles from a prior wet clean step. Another high risk is if a salicide process is done before the tungsten deposition. After the silicidation of the metal the excess metal needs to be removed. This is normally done in a wet metal strip step. One can understand that if metal residues are left at the dielectric layer, these can act as a catalyst for tungsten deposition. Especially with a Pt-salicide process this is known to be a severe problem.

With regards to this it is clear that the wafer pretreatment is of key importance for obtaining good selectivity. Such pretreatments can vary from wet clean steps to in situ dry clean steps. In situ clean steps have the advantage that they can be done in vacuo in an integrated (cluster) tool. Unfortunately not much has been published in the literature about the in situ pretreatments. In one study a NF_3 plasma is reported to be able to remove native oxide from silicon [Kajiyana et al.[84]].

Selectivity loss caused by the reaction products: The literature shows some evidence that reaction products such as HF and WF_x (in the hydrogen case) and SiF_x and WF_x (in the silane and displacement case) can cause loss of selectivity [Pauleau et al.[4], Blewer[122], Lami et al.[123], Kwakman et al.[124], McConica et al.[125], Sumiya et al.[126], Hirase et al.[127], Creighton et al.[128,129], Foster et al.[130], Broadbent et al.[131], McConica et al.[132]]. For instance, a reaction by-product like SiF_2 can after desorption from the tungsten adsorb again, but now on the oxide. A reaction like:

$$WF_6 \ + \ 3SiF_2 \ \text{-------} > \ W + 3SiF_4 \tag{3.11}$$

can proceed and cause loss of selectivity. Kwakman et al.[124], describe an interesting experiment wherein non-patterned wafers were sitting opposite either a silicon or a tungsten wafer in the reactor. The selectivity loss (H_2/WF_6 chemistry) was much more severe on the oxide wafers opposite the silicon wafers than on the oxide wafers opposite the tungsten wafers. This strongly suggests that SiF_x type of species are playing a role in the selectivity loss. By comparing the selective behavior of the H_2 and the SiH_4 chemistry Chow et al.[67], came to a similar conclusion. Undoubtedly the generation of silicon subfluorides will be much more pronounced in the case of the silane chemistry as compared to the hydrogen chemistry. In the latter case silicon subfluorides can only be generated in the beginning of the deposition in contact holes (to silicon). In the case of vias (to metal) no silicon fluorides can be generated. Thus in this case, selectivity loss due to silicon subfluorides can be excluded. In the silane chemistry situation, the risk of selectivity loss is higher when more silane is added to the gas mixture. This can be explained by the assumption that more SiF_x species will be generated with a lower x number. The lower x compound is a more powerful reducing agent, thus it will cause a higher probability for loss of selectivity.

Creighton[128,129], found the relatively stable WF_4 compound as a possible by-product which could, after adsorption on the oxide, initiate tungsten growth on oxide according to:

$$3WF_4 \ \text{-------} > \ W \ + \ 2WF_6 \tag{3.12}$$

Once the tungsten is formed at the oxide surface it will catalyze further tungsten growth.

It has been argued by Kwakman et al. 1988, that HF can cause selectivity loss according to:

$$HF \quad \text{------>} \quad F_a \quad + \quad H_a \tag{3.13}$$

or

$$SiF_{x,a} \quad + \quad (4\text{-}x)HF \quad \text{------>} \quad SiF_4 \quad + \quad (4x)H_a \tag{3.14}$$

The formed (activated) hydrogen can thus react with WF_6 to form tungsten.

We arrive to the conclusion that in order to maximize selectivity the generation of reaction by-products needs to be as small as possible. The most obvious way to achieve this is to make the hot surface in the reactor as small as possible. This is the driving force for using cold wall reactors for selective tungsten. Such reactors can show superior selectivity when compared to hot wall batch systems.

Local selectivity loss (creep up): One difficulty which occurs, especially with the silane selective chemistry, is the local loss of selectivity called "creep up". Creep up is the phenomenon that the tungsten growth starts not only from the bottom of the contact/via but also from the side walls (see figure 3.10). Creep up can cause problems such as cavity formation in the plug or shorts between contact or vias. The problem is very much more pronounced in vias to aluminum than in contacts to silicon. A possible explanation for this phenomenon is that during via etching the aluminum is redeposited at the side walls by the sputter action of the RIE etch. This can initiate tungsten growth at the wall of the via. Since the reaction runs in a depletion mode, enhanced deposition at the via mouth can occur and unwanted void formation is likely. Vias ending on materials such as TiW or W don't exhibit the creep up problem (see figure 3.12). The sputter etch yield for aluminum is about 2-3 times higher than that for tungsten [Glang et al.[253]]. Thus, the redeposition at the side walls will be less for TiW or W than for Al giving less rise to creep up phenomena for TiW or W.

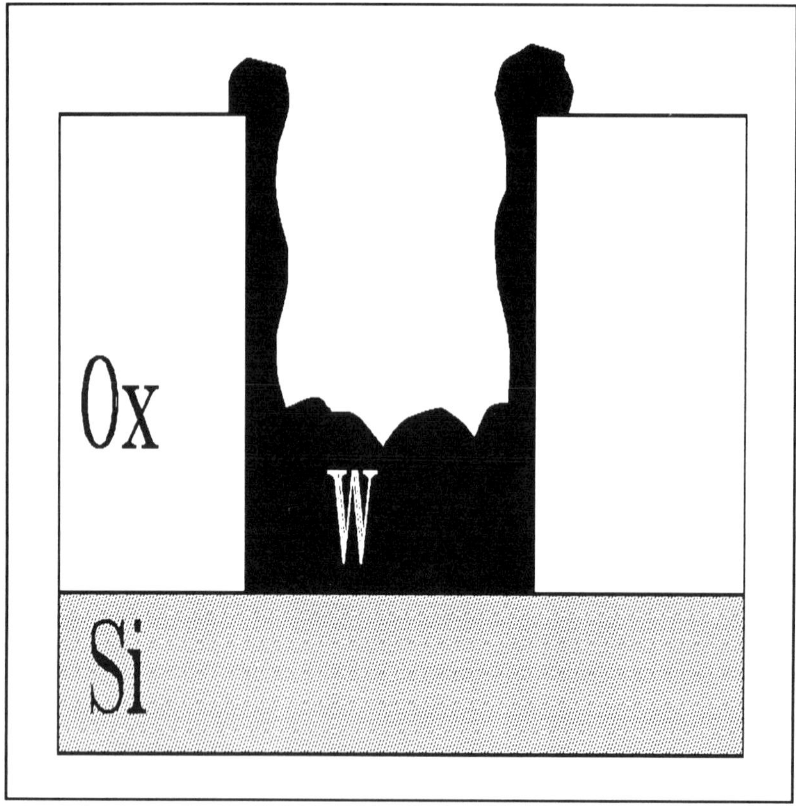

Figure 3.10. Local loss of selectivity: creep up.

Selectivity loss dependence on dielectric type: As mentioned in section 3.3 the use of many types of dielectric materials is possible. The order of selectivity loss for different dielectrics in terms of nuclei density has been studied for the hydrogen chemistry [Bradbury et al.[134]] and for the silane chemistry [Chow et al.[67]]. For both chemistries essentially the same order was found, where the best selectivity is on BPSG (or more generally doped oxides):

Doped Oxides > Undoped oxides (LTO, TEOS) > LPCVD-Nitride

This fact raises two important questions:

1) What is the reason for the observed order in selectivity?

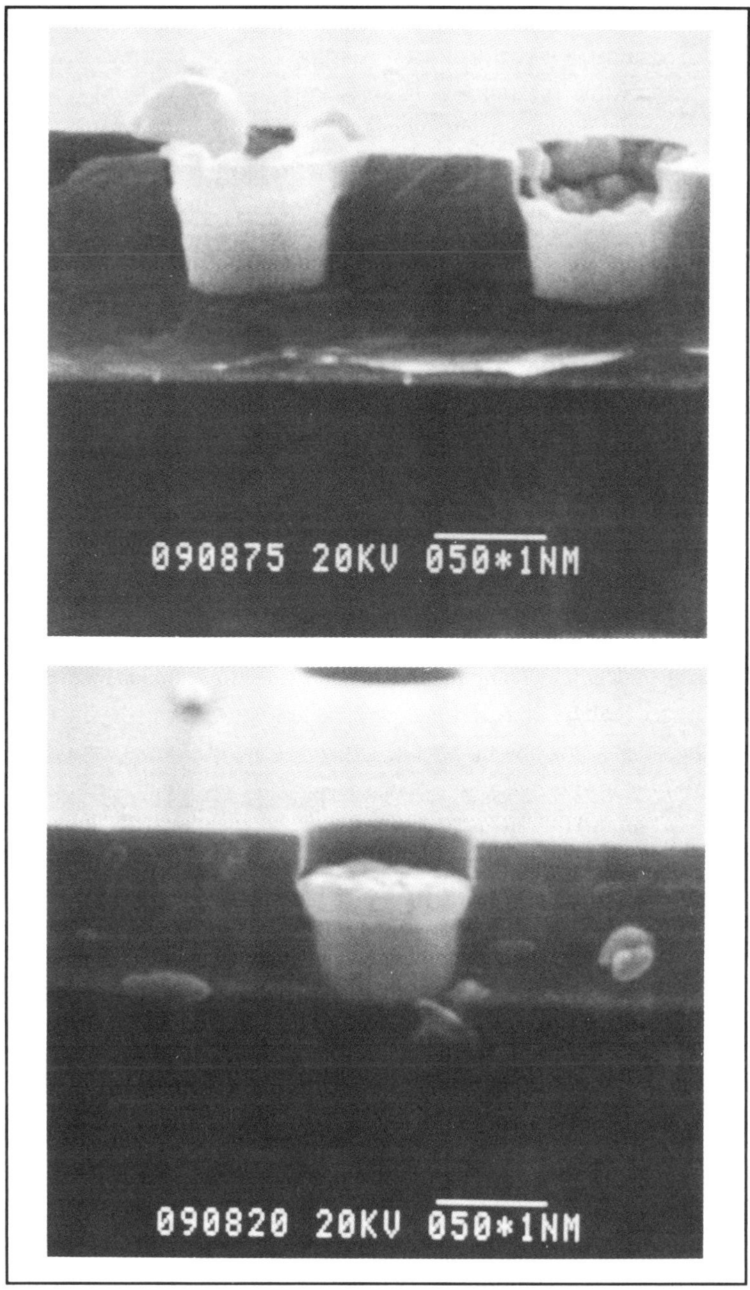

Figure 3.11. Difference in creep up: via to Al (top) and to TiW (bottom). [Courtesy of R. Chow, Genus, Inc.)

2) Does the fact that the same order is found for both chemistries give additional information relating to what reaction product is responsible for the loss in selectivity?

Let's start with the last question and have a look at what reaction products have been reported. Below 400°C the main Si containing species is SiF_4 in the SiH_4 chemistry [Yu et al.[29], Sivaram et al.[30]]. SiF_4 can be considered as not being chemically active since it can not carry more fluorine. Therefore, we don't expect SiF_4 as being responsible for any loss of selectivity. This has been confirmed experimentally (albeit only for the H_2/WF_6 chemistry) [Levy et al.[76]]. However, the experiments by Kwakman et al.[124] show clearly that silicon subfluorides play an important role in the selectivity loss. Thus the spectroscopic studies of Yu and Sivaram may not detect short living intermediates.

In both the silane and the hydrogen case we can safely assume that WF_x species will be among the reaction products. Once adsorbed at the dielectric surface, the sub-fluorides can undergo further reduction and will form tungsten nuclei. The tungsten nuclei will cause selectivity loss. (It is important to realize that HF is not a common reaction product for both chemistries since this has not been detected for the silane case but it is the major product in the hydrogen case!). Thus if tungsten sub-fluorides are at least partly responsible for selectivity loss in both chemistries we can expect the same trend for the silane and the hydrogen case with respect to the dielectric order for selectivity loss. As mentioned above a similar order for both chemistries is indeed observed.

An attempt to explain selectivity loss dependence on dielectric type has been proposed by Itoh et al.[85]. The assumption was that WF_6 has to adsorb to the surface prior to further reaction. In the WF_6 molecule the tungsten atom is hexagonally coordinated by 6 fluorine atoms. Thus it is envisioned that the interaction with the surface will occur via a fluorine atom. Of all elements fluorine has the highest electronegativity. Therefore, the interaction of WF_6 with the surface will be better the more electropositive the atoms in the substrate. The electronegativity (or electroposivity) of the substrate can be expressed by the Mullikan electronegativity scale. The lower the electronegativity of the substrate the better the interaction with WF_6 and the higher the risk for loss of selectivity.

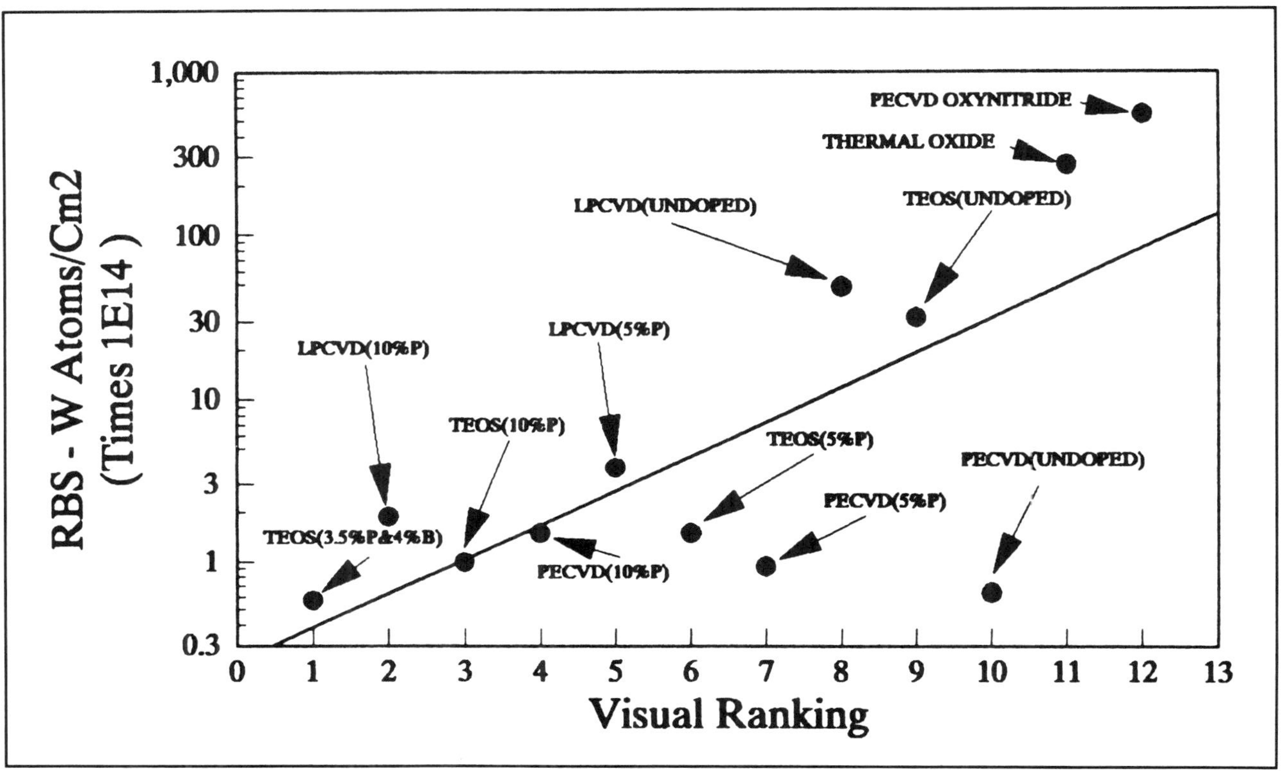

Figure 3.12. Ranking of the selectivity of SiH_4/WF_6 chemistry for different dielectric films. [Chow et al.[67], reprinted by permission].

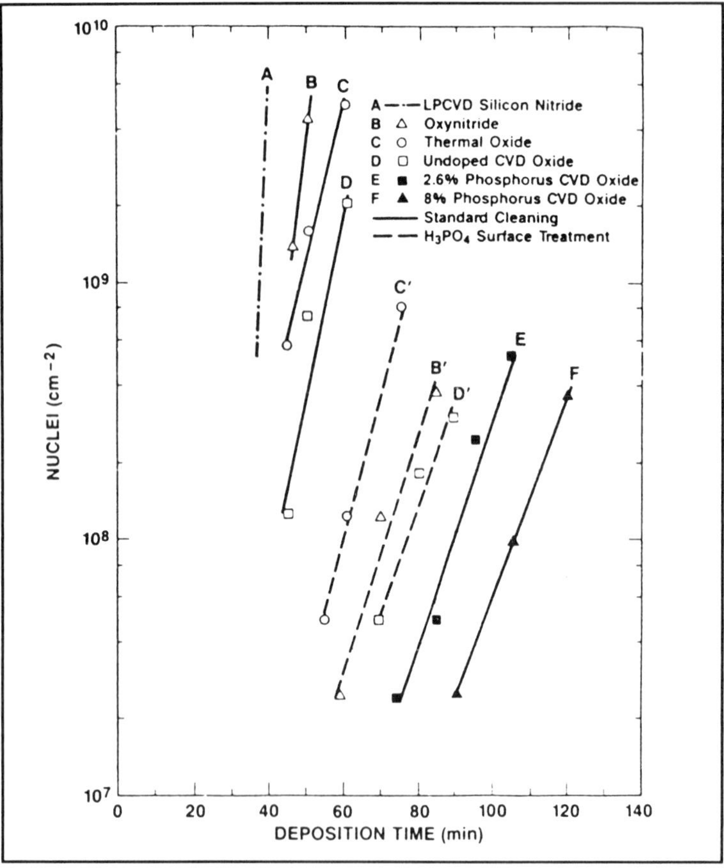

Figure 3.13. Ranking of selectivity for the H$_2$/WF$_6$ chemistry for dielectric films. [From Bradbury et al.[134], reprinted by permission of the publisher, The Electrochemical Society]

The order in electronegativity found by Itoh et al. was:

$$\text{Al-Al} < \text{Si-Si} < \text{Al}_2\text{O}_3 < \text{Si-N} < \text{Si-O} < \text{W-F}$$

On these substrate tungsten depositions were performed. The amount of selectivity loss followed indeed the trend predicted by the Mullikan scale. However, the doping of SiO_2 with B and P (which are less electronegative than O) is predicted to exhibit degradation in selectivity compared with undoped oxide. This is obviously not in agreement with the experimental evidence mentioned above (see also figure 3.12 and 3.13). Clearly more work remains to be done to explain all experiments consistently.

3.6 ELECTRICAL CHARACTERIZATION

Contact resistance and leakage current: With respect to contact resistance, there is a fundamental difference between the introduction of selective or blanket tungsten in an existing IC process (where no contact fill was applied). In most cases there will already be a barrier material present for contact reliability. Using a contact barrier layer, aluminum spiking and silicon precipitation from the AlSi at the Al/Si interface can be eliminated. This barrier material will more than likely be either sputtered TiW or TiN. In the case of blanket tungsten this same barrier layer can now be used as the glue layer. Since the specific contact resistance between the diffused or doped Si areas and the barrier layer is much greater than that between the barrier layer and the CVD-W (see figure 3.14), there will be essentially no change in the overall contact resistance of the contact upon the introduction of the blanket plug. This has been verified experimentally by Ellwanger et al.[7]. However, in the case of selective tungsten, the situation is dramatically different. Obviously the former barrier layer can no longer be used. The result is that a new, unknown electrically active interface is formed, namely that between the doped Si and the CVD-W. Therefore, much attention has been paid in the research of the selective tungsten process to characterize the contact resistance.

Another important phenomenon to be checked is the leakage current of shallow junction diodes. The leakage current provides valuable information about the quality of the Si-W interface and whether unallowed amounts of silicon are consumed during the selective tungsten deposition process. Again, one can expect that blanket tungsten gives less problems here because of the presence of the adhesion-barrier layer.

When characterizing such electrical parameters, it is of utmost importance to do the evaluation as accurately as possible and to use the appropriate control experiments. Some examples of what errors can be made

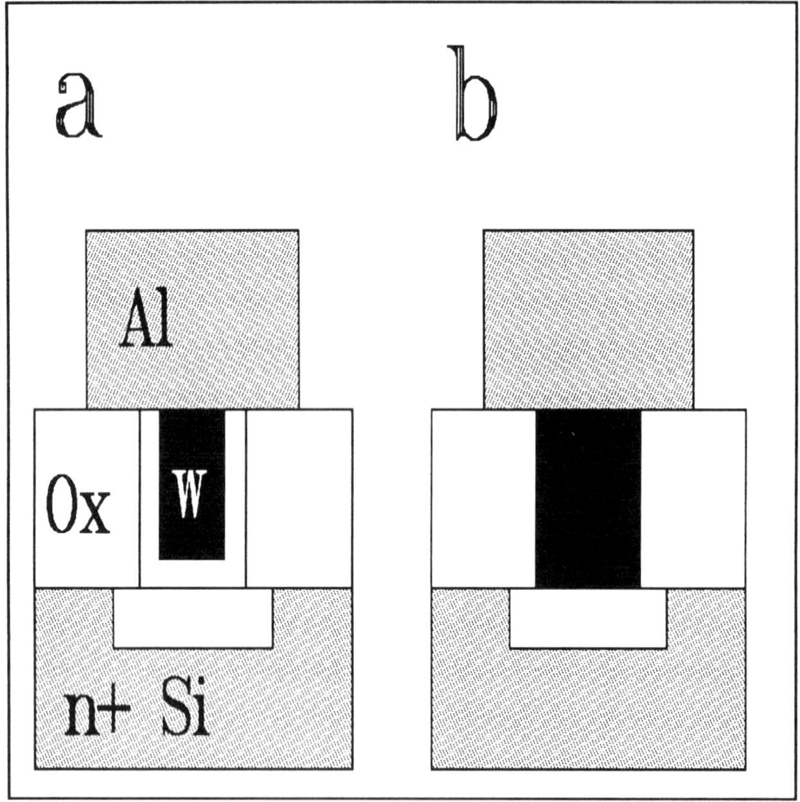

Figure 3.14. R_c is in the blanket tungsten case (a) determined by the glue layer/Si interface whereas in the selective case (b) it is the CVD-W/Si interface.

in an evaluation are:

i) After the contact resistance measurements are made, it is absolutely necessary to check the actual contact diameter and to verify that there was no over etching of the contact. In the case of overetching, R_c can be erroneously low because more contact area becomes available.

ii) Complications like encroachment can give also false readings since both the upper silicon layer (rich in dopant) can be consumed (giving a higher contact resistance) as a larger contact area can be formed (giving a lower contact resistance).

iii) For the evaluation of leakage current, appropriate diodes should be used. It should be stated what the values of junction depths are. Since leakage can occur preferentially along the perimeter of the diode instead of bulk leakage, it is of importance to design the experiment such that it is possible to discriminate between different modes of leakage. An interesting approach has been described in the literature [Ellwanger et al.[14]], see figure 3.15. We see two types of diodes: one type with almost no perimeter and one type with a very long perimeter. Also, one should fabricate the diodes such that the leakage current of the control group is low enough in order to be able to notice significant increases. As a rule of the thumb the leakage current of the control group should be of the order of 1 $\mu A/cm^2$ or lower.

The contact resistances reported [Levi et al.[76], Saraswat et al.[101], Tsutsumi et al.[102]] are consistently higher for W to p^+ silicon than for W to n^+ silicon which cannot be explained by elementary contact resistance theory [Levy et al.[76]]. This theory would predict, in the surface dopant range of about 10^{20} cm^{-3}, a comparable contact resistance for n+ and p+ silicon. The formation of an interfacial film (WF_4) with different barrier properties has been proposed [Levi et al. 1986] to account for this general observation that R_c to p^+- Si is higher than to n^+-Si. Another and more likely explanation has been proposed by Cohen[103]. He showed that the boron dopant can easily be deactivated by hydrogen. Hydrogen can form B-H bonds and then the boron is no longer electrically active leading to a high contact resistance on p+ silicon. Especially in the H_2/WF_6 chemistry such problems can be expected since excessive amounts of atomic hydrogen will be generated during the reaction. One way to overcome these problems is to clad the active diffusions with a silicide, for instance PtSi. Using PtSi specific contact resistances to Si in the range of 10^{-7} Ohms cm^2 for both n^+ and p^+ diffusions are reported [Levi et al.[76]].

The stability of the tungsten-silicon interface with respect to silicon diffusion and electrical integrity has been reported to be in the range 450-600°C [Joshi et al.[107], Shioya et al.[108], Pauleau et al.[109], Thomas et al.[110]]. This is very compatible with Al alloy temperatures (400-450°C) in VLSI processes.

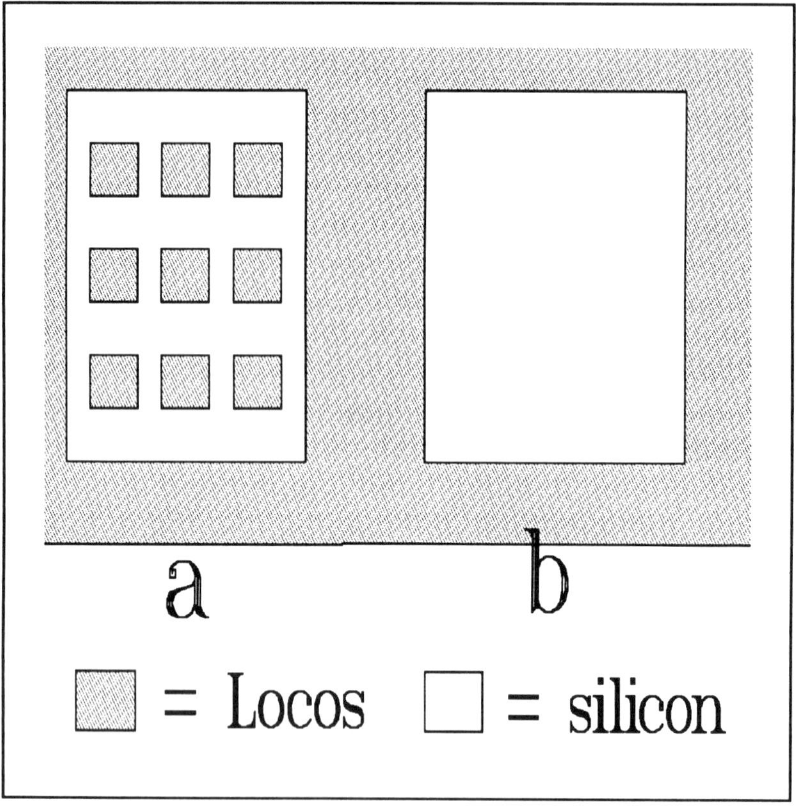

Figure 3.15. Top view of two types of diodes. Type a has much more perimeter and corners than type b.

The compatibility of the H_2/WF_6 chemistry with silicides is problematic for $TiSi_2$, PtNiSi [Broadbent et al.[111]] and $CoSi_2$ [van de Putte et al.[112]]. Problems such as Si extraction and void formation have been observed. In contrast, the milder SiH_4/WF_6 chemistry shows good results on such silicides [Ellwanger et al.[14]].

Contact reliability: When doing reliability studies one has to take care that the sample size (ie. the number of contacts or vias) is sufficiently large in order to come to statistically meaningful conclusions. The normal way to do this is by using chains in which thousands of contacts or vias are connected. For contact chains or strings this can, depending on how exactly the diffusions are formed, lead to a very high string resistance. In order to achieve enough current density, necessary to stress the contact, high voltages are required. This voltage drop, if high enough, can cause junction break down. Therefore, in most cases via strings are studied where these problems do not occur. One should be alert to the fact that, since the current distribution in a contact can be quite different than in a via, the MTF (mean time to failure) figures obtained using via strings don't necessarily coincide with those of real contacts.

A key question here is how much will the contact or via reliability improve by incorporating tungsten plugs. It is important to realize that the MTF of a via string can never be better than the MTF of a metal line of comparable length without vias. In other words, the overall reliability also depends on the type of interconnect material. For instance, the electromigration resistance of AlSiCu alloys are known to be much better than that of AlSi alloys. Needless to say that the test structures should be designed such that upon testing, the stress is really at the plug and not in the interconnect material. Unfortunately not much has been published on this subject [Itoh et al.[104], Nordstrom et al.[105], Matsuoka et al.[106]]. Work that has been reported was done on plugs which were formed by a selective tungsten process. It is not necessarily true that results obtained for selective plugs will also be applicable for plugs formed by a blanket tungsten process.

In a recent elegant study by Matsuoka et al.[106] came to the following conclusions:

i) For tungsten filled vias there is no dependence of the MTF on the via diameter. This is in contrast to the conventional aluminum filled vias where a strong dependence is on the via size (see figure 3.16).

ii) The MTF for the tungsten filled vias is about 50 times that of the conventionally filled vias (1um diameter).

iii) The activation energy of the MTF for the plugged vias is 0.62 eV. This in the range of aluminum grain boundary diffusion (0.5-0.6 eV) which suggests that aluminum electromigration at the plugs is the failure mode.

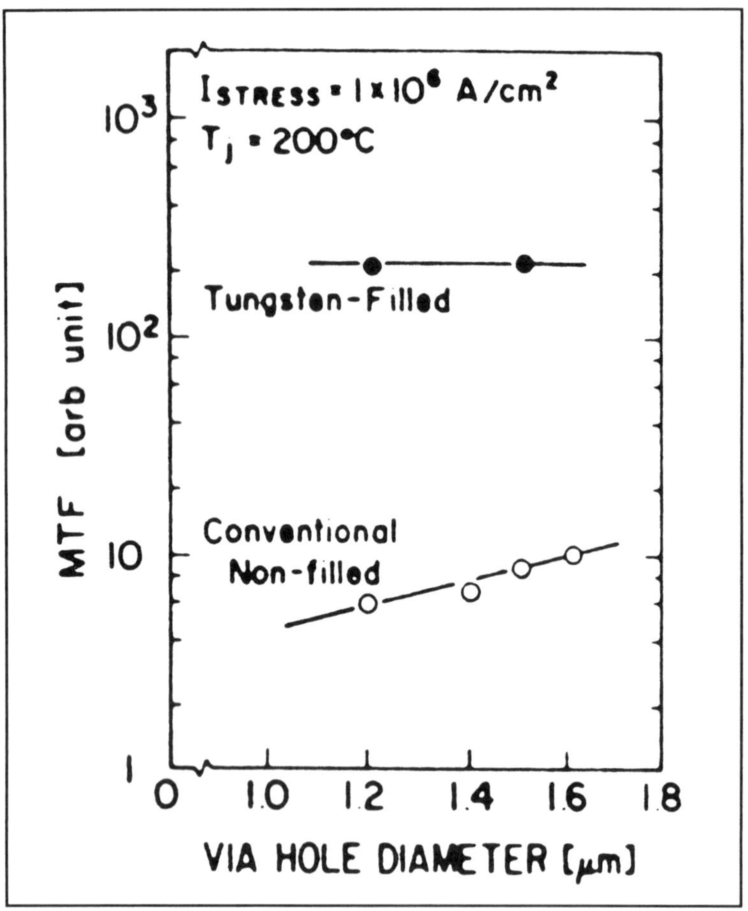

Figure 3.16. MTF of tungsten filled and conventional filled vias. [Matsuoka et al.[106], reprinted with permission, © 1990 IEEE].

iv) During current passage under stress conditions (200°C, 1×10^6 A/cm^2) there is an increase in via contact resistance. It was shown that this was due to Si precipitation at the anodic side of the plug-Al interface. Under normal operation (125°C, 3×10^5 A/cm^2) this will cause no problems (0.05 Ohm increase per via in 10 years).

CHAPTER IV

BLANKET VERSUS SELECTIVE TUNGSTEN

The main application of CVD-W in the immediate future is contact or via fill. We have seen that both selective and blanket tungsten can give plugged contacts and vias. Therefore, we need to investigate what process will be the first choice for a given situation. Two aspects are important: the feasibility and the costs of the contact/via fill process. In addition, a very important criteria will be at what time will the fill process be needed in production. Let us first focus on the feasibility and process requirement aspects.

4.1 FEASIBILITY OF SELECTIVE AND BLANKET CONTACT OR VIA FILL

In table 4.1 a comparison is made between selective and blanket tungsten for contact and via fill. In the following we will discuss each of these process requirements.

Contact Diameter: In principle, selective tungsten has almost no limitations as to the contact size. However, especially in the SiH_4/WF_6 case, the local growth rate can drop when the contact size is too large, when the contact density is very high, or when the scribe lines are open. Blanket tungsten has, as discussed in chapter II, an upper limit for the contact size.

TABLE 4.1

Process limitations for blanket and selective contact/via fill

Process limitation	Selective	Blanket
Contact diameter	no	yes
Contact depth	yes	no
DOP[*]	no	yes
Compat. with substrate	no	yes
Effect exposed active area	yes	no
Interconnect option	no[**]	yes

[*] DOP = degree of planarization
[**] However, see ref. 156 for the encapsulation of Al lines

Contact Depth: Selective tungsten has a fundamental problem of simultaneously filling contacts of different depths (see below). There will always be contacts with either an overfill or an underfill (see figure 4.1). Overfill leads to loss of real estate or yield and underfill can give step coverage problems for the aluminum. Blanket tungsten has almost no limitation here (as long as the step coverage is sufficient, and contact diameters are below the limit dictated by the film thickness). Since in the case of the vias the depth variation is much less, selective tungsten will be more appropriate for via fill.

Degree of Planarization (DOP): Whereas the result of the blanket tungsten etch back will be very sensitive for the DOP, selective tungsten is virtually

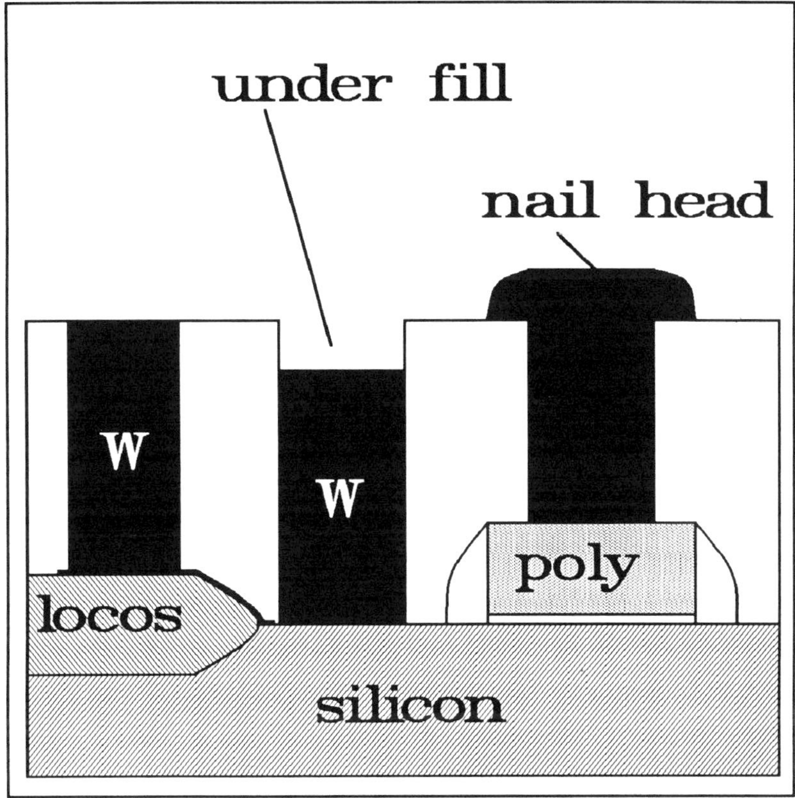

Figure 4.1. With a high DOP contacts will have variable depth. This gives a fundamental problem for selective tungsten contact fill.

unaffected by the DOP (however, planarization gives variation in contact depth, see above). As stated in chapter II, however, planarization in VLSI/ULSI devices is necessary for other reasons than just blanket tungsten.

Compatibility with The Substrate and Dielectric: The adhesion layer in the blanket-W case now proves to be an advantage. Most commonly this is sputtered TiW or TiN. The contact resistance and reliability issues of these materials have been well studied and are under control. Thus for blanket tungsten no additional problems are expected here. In the case of selective tungsten a completely new interface is created, namely that between silicon and the tungsten. Also, because of the wide variety of substrates and dielectrics (see table 4.2) it is almost impossible to fulfill all needs with one generic selective tungsten process. In principle for each dot in the matrix one has to find new process optimums. This fact is the main reason for the slow progress made with the incorporation of selective tungsten in pilot production. When we look to via fill only, there is a severe reduction in the different combinations. Unfortunately the inter-metal dielectrics (plasma oxide, plasma nitride etc.) typically yield poorer results in terms of selectivity as compared to the dielectrics used at the contact level (see chapter III). A compensation for this difficulty might be that one does not need to worry about the leakage current behavior but only the contact resistance of the via. This makes optimization of a selective process at the via level less cumbersome.

Effective Exposed Area: This aspect is especially important for the silane based selective deposition which runs in a reactant feed controlled regime. This implies that the <u>local</u> growth rate can depend on the amount of exposed active area. For instance, larger contacts can fill with a slower rate than small contacts. Or, when the scribe lines are exposed this can slow down the overall growth rate in the contacts and this effect is indeed observed [Chow[267]]. In other words, the deposition rate is not a constant but merely depends on the given environment. Blanket tungsten clearly will not suffer from such effects.

Interconnect Option: A very attractive feature of blanket tungsten is that the fill step can be combined with an interconnect option. Several successful examples are mentioned in the literature (see for a brief excerpt section 5.2). Selective tungsten cannot offer this option.

4.2 COSTS OF THE CONTACT/VIA FILL PROCESS

The total costs of the fill step depends strongly upon the amount of process steps and the amount of WF_6 needed. Although it is widely believed that selective tungsten contact fill has fewer process steps as compared to

90

Blanket versus selective tungsten.

TABLE 4.2

CONTACT VIA FILL USING SELECTIVE W

DIELECTRICS

SUBSTRATES	BPSG	LTO	TEOS	PETEOS	PE-OXIDE	PE-NITRIDE	SOG	POLYIMIDE
mono-Si	•		•	•			•	
poly-Si	•		•	•			•	
PtSi	•		•	•				
TiSi$_2$	(•)		•	•			•	
CoSi$_2$	(•)		•	•			•	
TiW		•			•	•	•	•
TiN		•			•	•	•	•
Mo		•		•	•	•	•	•
Al		•		•	•	•	•	•

• = a possible contact or via/dielectric combination

91

blanket tungsten, this might not be true. See table 4.3.

TABLE 4.3

Overview process steps selective/blanket tungsten

Step	Selective	Blanket
1	in situ wafer pretr.	in situ adh. layer
2	selective deposition	blanket deposition
3	post depos. treatment	blanket etch back

The post deposition treatment step in case of selective tungsten might be necessary to restore the effects due to selectivity loss or to remove the nail heads formed due to the depth variations of the contacts. We see that the amount of process steps (and the associated yield losses) can well be equal for selective and blanket tungsten. Fortunately, in both cases the three process steps can in principle be integrated in a cluster tool which can improve repeatability (see chapter VII).

In terms of process costs the pertinent issue is the high consumption of WF_6 in the blanket tungsten case: about 300 scc/per wafer for blanket and about 30 scc/per wafer for selective tungsten (one micron film thickness). Also, pure WF_6 (99.999%) is rather expensive. Therefore, it is attractive to explore ways of reducing the amount of WF_6 needed by proper reactor design and to investigate lower grades of WF_6 in the case of blanket tungsten. Nevertheless, at this time blanket tungsten contact fill is still regarded as an expensive process step. This will be counter balanced when higher yields are obtained and better reliability performance of circuits is demonstrated.

4.3 WORLD WIDE STATUS OF CVD TUNGSTEN

Blanket tungsten for contact/via fill and interconnect applications has been accepted mainly in the USA and Europe. Many IC manufacturers have plug processes in pilot production and are moving on to large scale production. In Japan and the Far East a lot of effort was, and still is, focused on selective tungsten. However, due to the complexity of selective tungsten process control (certainly in contacts), increased interest from these areas in blanket tungsten can be observed. It is expected that in the next two years the acceptance of the blanket tungsten approach will become more pervasive worldwide.

Although selective tungsten is proposed and extensively studied as the first candidate for contact fill, only one (Japanese) or two companies have reported implementing such a process in pilot production. Thus, the "real live" pilot line experience with selective tungsten is very limited as compared to blanket tungsten.

4.4 CONCLUSIONS

Based on the experience so far, blanket tungsten is the only **production proven** solution for contact or via fill. It is expected that rapid general acceptance in the next two years will occur. This implies that it took blanket W plug and interconnect technology about 5 years to become accepted which is surprisingly fast.

The first opportunity of selective tungsten to become incorporated in IC manufacturing will be for a via fill application. In order to achieve this a very close and extensive cooperation between equipment vendors and major IC manufacturers is essential. Only in this way will a production compatible process be developed.

In cases where tungsten interconnect is possible, blanket tungsten will remain the most attractive technology. The only replacement for this process could be a copper deposition (CVD or electroless) process. Indeed, several preliminary studies in this field have already been reported [Arita[151], Kelber et al.[152], Hazuki et al.[153], Pai et al.[154], Hu et al. 1990[155]].

CHAPTER V

TUNGSTEN AS INTERCONNECT MATERIAL

5.1 WEAKNESSES OF ALUMINUM INTERCONNECTS

Until now aluminum has been the first choice material for interconnect applications. This has mainly been due to properties such as:

- Low resistivity. The bulk resistivity of pure aluminum is 2.74 $\mu\Omega$ cm. Only the metals gold, copper and silver have a lower bulk resistivity.

- Etchability. Aluminum can be easily etched with good etch selectivity towards oxides.

- Deposition. Aluminum can be deposited using the sputter technique, a process which is well characterized and manufacturable.

With the increasing amount of integration, however, some fundamental problems arise with the use of aluminum, such as:

- Electromigration. Smaller geometries lead to higher current densities which give more reliability constraints. Although the electromigration characteristics of aluminum can be improved by proper doping of aluminum with elements such as copper [D'Heurle[91]], other additional problems in the area of etching and corrosion may be introduced.

- Corrosion. Aluminum is a not a noble metal and is attacked by both alkali and acidic solutions. Because of the presence of a surface Al_2O_3 film, the metal is protected against corrosion [Diggle et al.[136], Borgmann et al.[137]]. This oxide film, however, is easily penetrated, for instance, by the presence of chlorine ions which remain in the resist after a chlorine based plasma etch. Also, the presence of Cu in the aluminum weakens the corrosion resistance of the alloy by the presence of an unfavorable electrochemical couple (Al/Cu^{2+}).

- Stress related phenomena such as void and hillock formation. Due to the action of the dielectric stress, void formation can be induced [Hinode at al.[138]]. Hillocks can be formed due to intrinsic or thermal stress [Gardner et al.[139]).

- Si (epitaxial) precipitation in contact areas to silicon [Hirashita et al.[140]). The main problem here is an increase in contact resistance, especially in small contacts, thus leading to device reliability problems.

- Step coverage. Due to the poor step coverage properties of the PVD technique, the reliability of the aluminum metallization suffers from opens due to enhanced electromigration in the contacts and vias during current passage (see chapter II).

Since several of these problems are inherent to the use of aluminum, other materials have been investigated. An increasing interest, as indicated in the literature, to the application of blanket tungsten as the interconnect material because of its higher resistance against electromigration [Kaanta et al.[142]] has been shown. The main drawback with tungsten is its higher bulk resistivity compared to that of Al(Cu,Si) (5.3 versus 3.5 $\mu\Omega$cm respectively). This can cause severe problems for high speed devices. Nevertheless, there is some belief that this problem can be overcome by a proper design of the interconnect system, although it may put constraints on circuit design. In the following section we will discuss some recent results obtained with tungsten as an interconnect as described in the literature.

5.2 TUNGSTEN INTERCONNECTS

Since 1987 several papers have described the incorporation of blanket tungsten CVD for interconnect applications.

Kaanta et al.[142] implemented tungsten for contact fill, via studs, and the first interconnect metal. The via plug was produced by the blanket etch back method. The apparent disadvantage of the higher resistivity of tungsten as the interconnect was compensated by:

a) Using a thinner tungsten film which gave a RC gain due to less capacitance. The resulting higher current density is not a problem since tungsten has a good electromigration resistance compared to aluminum.

b) By using a thicker Al film for metal 2 inwhich long interconnects were designed.

The use of tungsten plugs allowed for a circuit density improvement of about 15% (see also the discussion in paragraph 1.4 about the effect of tungsten on the design rules). The tungsten interconnect was planarized using a combined deposition-sputter etch process. No failures were obtained during accelerated current stressing in the tungsten part of the circuits.

Chapman et al.[143] used W as the metal 1 material, they filled the vias using a selective tungsten process.

A 2 um pitch triple-level metal process using two levels of tungsten interconnect was reported by Bonifield et al.[144]. The third metal level was still Al-Cu(2%) driven by the requirements of wire bonding. For the vias between metal 2 and metal 3 a selective tungsten process was used.

Planarization of the tungsten lines was accomplished using the REB technique.

Arena et al.[147], reported CVD-tungsten for interconnect while using sputtered tungsten as the adhesion layer to BPSG-oxide.

Nakasaki et al.[146] addressed the tungsten-oxide adhesion problem using reactively sputtered TiN as adhesion layer. A minimum thickness of 100 Å TiN was needed for adhesion.

Brasington et al.[141] reported an interconnect scheme in which a stack of a 100 nm TiW glue layer, 450 nm CVD-W and 450 nm Al was used. The Al layer on top reduced the overall sheet resistance of the interconnect stack.

5.3 ISSUES OF TUNGSTEN INTERCONNECTS

There are at least four points of concern when using tungsten as the interconnect material. These are stress, roughness, resistivity, and etchability of the film. In the sections below we will discuss each of these problem areas.

5.3.1 Tungsten Film Stress

The stress of CVD-W films can vary, depending on the deposition conditions [Joshi et al.[51], Clark et al.[52], Blumenthal et al.[148], Sivaram et al.[149]), by one order of magnitude (ie. from $3x10^9$ to $13x10^9$ dyne/cm^2) and is mostly tensile. Experience has shown that for a plug process the stress is seldom a problem since the majority of the film is removed during the etch back process. Loss of adhesion is usually not observed in the blanket-plug process. When the interfaces between the different films are clean the adhesion will be formed by chemical bonds (1-2 eV) instead of (weak) physical forces (ca. 0.2 eV).To remove a film with an adhesion of 1 Ev per atom, a stress of about 20×10^9 dynes/cm^2 is needed [Campbell[150]]. Often in the case of peeling problems, the stress of the film is blamed. This is usually a wrong assumption.

For an interconnect application the situation will be different than for a plug application. Two problems can occur:

- After patterning, the tungsten lines can lift. In this case the situation is different than above since now we have discontinuities such as line edges, corners in tungsten lines, and the end of a tungsten line. Here the stress can locally be much higher, is non-isotropic and can lead to problems like enhanced etching of the adhesion layer.

- When the stress is too high, the curvature of the wafer can become such that problems in the lithographic stepper and other equipment arise. A 6" wafer on which a one micron thick film is deposited with a tensile stress of 10×10^9 dyne/cm^2, can have a bow of about 50 μm! (see figure 5.1). This can result in focusing problems in lithography equipment or clamping problems on vacuum chucks. Generally these problems will not occur when the stress is in the range of $5\text{-}7 \times 10^9$ dyne/cm^2.

5.3.2 Origin of the Stress in CVD Tungsten Films

The stress in a thin film is composed of three contributions [Campbell[150]]:

$$\sigma = \sigma_{external} + \sigma_{thermal} + \sigma_{intrinsic} \qquad (5.1)$$

where σ is the symbol for stress. The external applied stress is in our case non existent. The thermal stress originates because of a mismatch between the linear expansion coefficients of tungsten and silicon (see table 6.4). An estimate of the thermal film stress can be made by using expression 5.2:

$$\sigma_{thermal} = (\alpha_W - \alpha_{Si})[E_W/(1-\nu_W)](T_{dep} - T_{RT}) \qquad (5.2)$$

where α_W and α_{Si} are the linear expansion coefficients of tungsten and

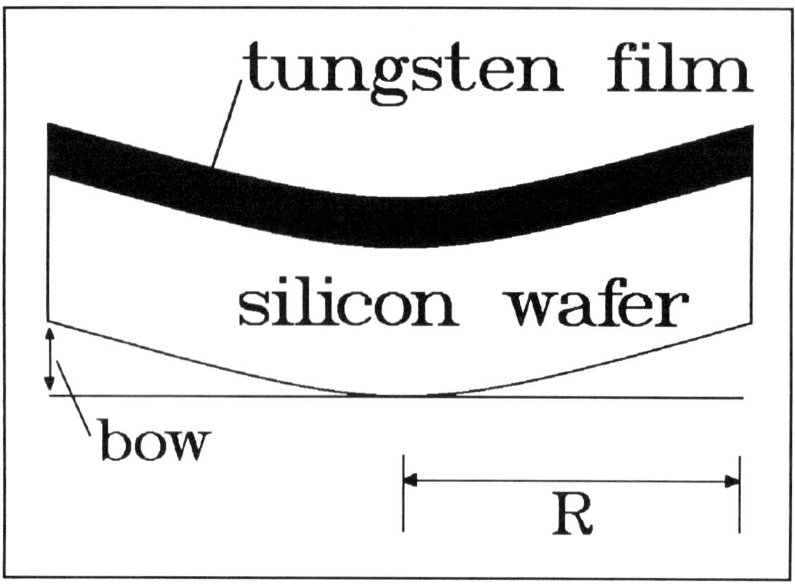

Figure 5.1 Bow of a wafer on which a one micron thick tungsten film is deposited with a tensile stress of 10×10^9 dyne/cm^2. A 6" wafer can have a bow of approximately 50 μm.

silicon respectively, E_W is the Young modulus of tungsten, ν_W is the Poisson ratio of tungsten, T_{dep} is the deposition temperature, and T_{RT} the room temperature. At a deposition temperature of 400°C and a room temperature of 20°C, expression 5.2 produces a stress of approximately 2.6×10^9 dynes/cm^2. (We assume here that the elastic modulus and the linear expansion coefficient of the poly-crystalline W film will be identical to that of bulk tungsten). Since the stress typically obtained for a film grown at 400°C is about 10×10^9 dynes/cm^2 tensile, we conclude that the intrinsic stress is of the order of 7×10^9 dynes/cm^2.

Several studies showed that the intrinsic stress of the CVD-W films could be reduced by going to a higher deposition temperature. According to recent studies [Yoshi et al.[51], Clark et al.[52]] of high pressure H_2/WF_6 chemistry, low stress films can be obtained at deposition temperatures of 470°C and higher (see figure 5.2). In this temperature range the stress also depends on the WF_6 partial pressure, lower WF_6 pressures give lower tensile stress. Unfortunately, these conditions (i.e. low WF_6 flow and high deposition

100

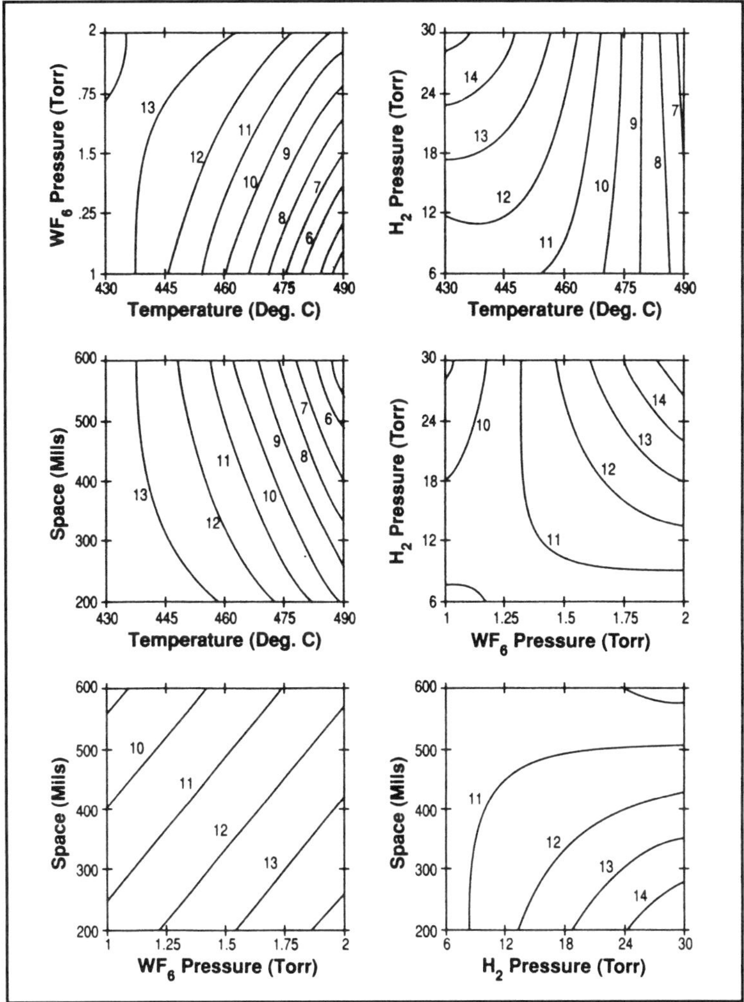

Figure 5.2. Dependence of tensile stress (10^9 dyne/cm^2) of CVD-W films on process parameters, H_2/WF$_6$ chemistry. [Clark[52], reprinted with permission, © 1990, Lake Publishing Corporation].

temperatures) do degrade the step coverage as can be inferred from equation 2.11. A solution suggested by Clark et al. is a two step deposition process. In the first step a high partial pressure WF$_6$ process is used to yield reasonable step coverage although relatively high stress. Here the thickness is chosen such that the contacts will be filled. In the second step the WF_6 partial pressure is reduced and the stress is thus decreased. The composite film can now have an acceptable stress for the interconnect application.

5.3.3 Determination of Film Stress

The determination of tungsten film stress can be done in several ways. A convenient method is to determine the bow of the wafer and use the equation of Stoney:

$$\sigma = (t_{Si}^2/t_W)[E_{Si}/(1-\nu_{Si})](\Delta r/3R^2) \qquad (5.3)$$

where t_{Si} and t_W are the thickness of the silicon wafer and the tungsten film respectively, E_{Si} is the Young modulus of silicon, ν_{Si} is the poisson ratio of silicon, Δr is the change in the bow of the wafer due to the film, and R is the wafer radius. The bow of the wafer can be measured using dedicated equipment such as laser reflectometry or a stylus scan.

5.3.4 Roughness of Tungsten Films

Rough films can cause problems in two ways:

i) The lithographic equipment has difficulty in finding the alignment marks and

ii) Problems can arise with respect to the critical dimension control of the line width when patterning the material.

In both cases smoother films are desirable although this is not to say that the patterning of rough films is impossible. As examples, many of the studies mentioned in paragraph 5.3.2 probably use low pressure processes which yield rougher films.

The roughness of the tungsten film has been found to depend

strongly on the total pressure in the system [Joshi et al.[51], Clark et al.[52]] (see also chapter II). The reason for this behavior has not been explained but the answer is likely found in terms of a pressure dependent nucleation density. In figure 5.3 the reflectivity is plotted as a function of the process pressure. The rougher the film the lower the reflectivity (see below). Recently, Sakamoto et al.[264] have reported that small additions of N_2 to the gas mixture can improve the surface roughness further.

The determination of the surface topography or surface roughness of a thin film is not a trivial exercise as has been discussed by Verkerk and Raaijmakers[175]. The difficulty with a technique such as SEM is that translation into quantitative numbers is not straightforward. The stylus scan can give such numbers but because of the size of the stylus this might not be representative for the actual topography. Verkerk and Raaijmakers showed that light scattering is a suitable technique for characterization of surface topography. Kamins et al.[177] studied the relationship of the reflectivity and the surface roughness for CVD-W films. The principles of the method will be discussed below. It can be shown that the reflectivity for normal incidence is related to the r.m.s. roughness S_R by:

$$S_R = [\lambda/4\pi]\ [\ln(R_o/R)]^{1/2} \tag{5.4}$$

where S_R is the surface roughness in nm, R_o is the reflectivity of a specular tungsten film, R the reflectivity of the film and λ the wave length in nm at which the measurement is done. The equation holds only in the smooth surface limit which means that $S_R \ll \lambda$. A good correlation between the roughness calculated from equation 5.4 and the actual roughness as estimated from microscopy was found by Kamins et al.[177]. A graphical representation of equation 5.4 is given in figure 5.4. In the range of 20 to 80% reflectivity, the relation is almost linear. A practical problem might be the determination of R_o. One way to do this is to take a (CVD) tungsten sample with a known specular surface. This will not always be possible. Another way is to calculate R_o from the optical data (refractive index n and transmission coefficient k) of tungsten using the Fresnel equation:

$$R_o = [(n-1)^2 + k^2]/[(n+1)^2 + k^2] \tag{5.5}$$

At 436nm, n=3.31 and k=2.47 [Palik[176]] therefore, $R_{o,W} = 46\%$. Because

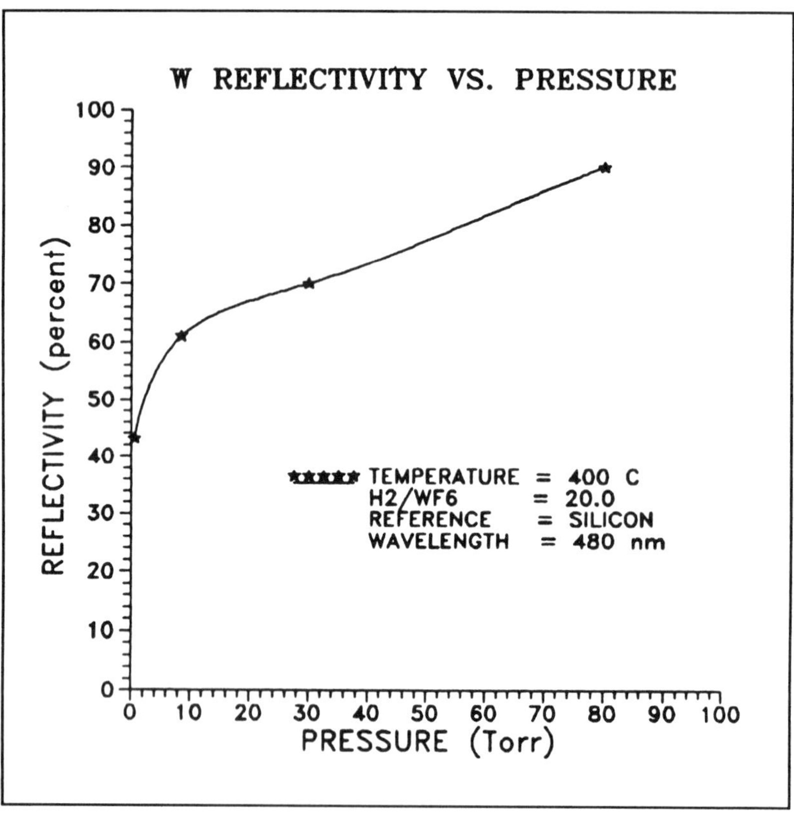

Figure 5.3 The reflectivity as a function of the total deposition pressure. The films are one micron thick. [Joshi et al.[51], reprinted with permission].

of the lack of a good specular standard it is common to relate the reflectance of CVD-W to that of mono-crystalline silicon and obtain a relative reflectance:

$$R_{rel,W} = R_W/R_{0,Si} \qquad (5.6)$$

In order to convert this number to the absolute reflectivity R_W, we need to know $R_{0,Si}$. This can be done using equation 5.5 again. The result at 436nm is $R_{0,Si} = 0.44$. Thus, as an example, if the reflectivity relative to silicon is measured to be 40% than the absolute reflectivity is 0.4x0.44=0.18. The ratio R_0/R in equation 5.4 becomes 0.46/0.18=2.6 and S_R can be calculated to be 34nm. Note that $R_{rel,W}$ can become larger than 100% since $46/44=(R_{0,W}/R_{0,Si})=105\%$.

5.3.5 Resistivity of CVD-W

The reported bulk resistivity (ρ) of tungsten films deposited with the H_2/WF_6 chemistry varies between 7 and 12 $\mu\Omega$cm. Compared with the bulk resistivity of sputtered Al(Cu), (ca. 4 $\mu\Omega$cm) this represents an increase of a factor 2 or 3. Since the line resistance can always be considered a parasitic resistance, it is important to keep the bulk resistivity as low as possible. Only a few reports have dealt specifically with the resistivity of blanket tungsten (mostly H_2/WF_6 chemistry). Before we review these reports, let us first briefly summarize some concepts of conductivity in metals [Maissel[178], Eckertova[179]].

The bulk resistivity is defined by:

$$R = \rho \ l/S \tag{5.7}$$

where R is the resistance between two contacts, l the distance between the contacts and S the cross sectional area of the film. See below for the determination of ρ using a four-point probe.

For bulk material, simple theory leads to the following equation:

$$\rho_o = N \ e^2 \lambda_o/mu \tag{5.8}$$

where ρ_o is the resistivity of the bulk material; N the electron concentration; e and m the electron charge and mass respectively, u the thermal mean velocity and λ_o is the electron mean free path. Basically it is assumed that the interaction of electrons with the perfect lattice does not contribute to the resistivity. Only collisions with imperfections like impurities, defects and thermal lattice vibrations (phonons) contribute to the resistivity. In the case of a poly-crystalline thin film the overall resistivity is composed of:

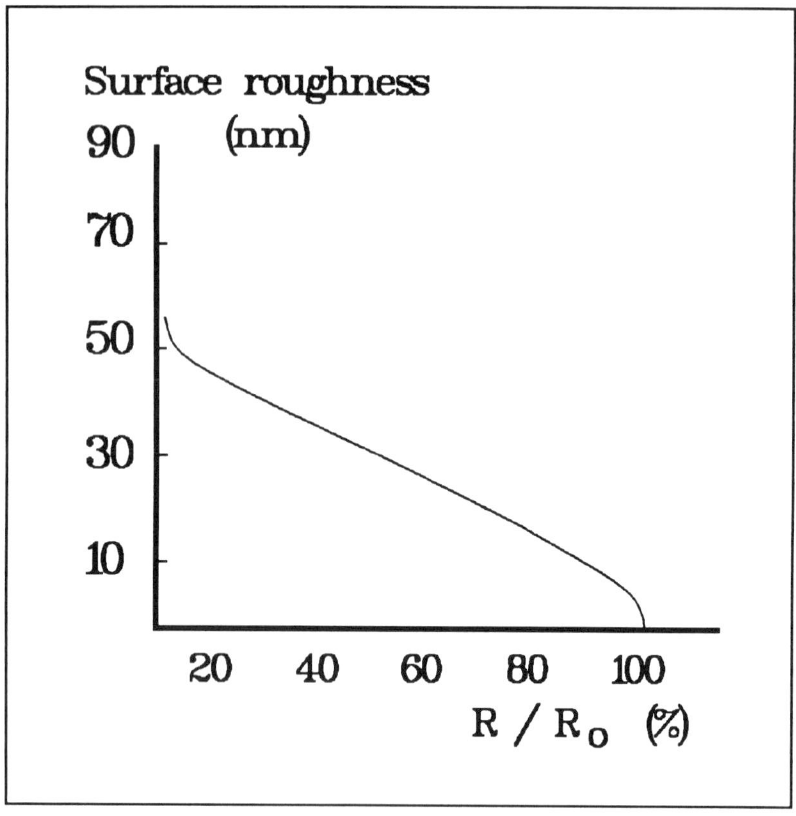

Figure 5.4 Roughness as a function of reflectivity as calculated with equation 5.4. The wavelength is 436nm.

$$\rho = \rho_{ph} + \rho_{irr} + \rho_{sb} + \rho_{gb} \tag{5.9}$$

where the subscripts ph, irr, sb and gb refer to the electron interaction with the phonons, with irregularities (impurities, defects), the surface boundaries and with grain boundaries, respectively. It is well known that the resistivity of a thin film is larger than that of the bulk material. Several theories exist to explain this phenomenon. Here we give only the results of an analysis based on the Boltzmann transport equation (single crystal material). For a film of thickness $t \gg \lambda_o$, the pertinent equation is:

$$\rho = \rho_o[1 + 3\lambda_o/8t]^{-1} \tag{5.10}$$

and for $t < \lambda_o$:

$$\rho = \rho_o(3t/4\lambda_o)[\ln(\lambda_o/t) + 0.4228] \qquad (5.11)$$

The mean free path for the electrons in single crystalline tungsten is about 40 nm. Equations 5.10 and 5.11 hold for diffuse reflections of the electrons at the surfaces of the film. In the case that a certain fraction, p, of the electrons reflect specularly at the surface boundaries, equation 5.10 will become:

$$\rho = \rho_o[1 - 3\lambda_o/8t]^{-1}(1-p) \qquad (5.12)$$

In the case that all reflections are specular, i.e. p=1, there will be no thin film effects on the resistivity. The theory shows further that for ratios t/λ_o > 5, hardly any thin film effect is predicted and the value of ρ should approach that of bulk tungsten. This is of course only true for structurally perfect (single crystalline) films.

It should be mentioned that the resistivity of tungsten is not very critical in the case of a contact fill. For instance, for a one micron diameter and one micron deep contact the total resistance of the plug (assuming uniform current density) is $1.3 \times 10^{-2} \times \rho$ Ω, where ρ is in $\mu\Omega$cm. We see that for a film resistivity of 10 $\mu\Omega$cm the total plug resistance is about 0.13 Ω. This value compared to the contact resistance, which will be of the order of 20 Ω, is a negligible amount. The situation is different for vias since a typical via contact resistance is of the order of 0.5 Ω. Nevertheless, it can be said that the resistivity of tungsten for contact plug applications can have values as high as 20-30 $\mu\Omega$cm.

The resistivity of CVD-W for both the H_2/WF_6 and the SiH_4/WF_6 chemistries has been the subject of several papers.

Learn and Foster[180] showed a strong dependence of the resistivity on film thickness for tungsten deposited at 400°C (see figure 5.5, H_2/WF_6 chemistry). They speculated that the detected 0.07 at% oxygen in the film might accumulate at the grain boundaries thus creating additional electron scattering. It was also found that the grain size decreased with thinner films,

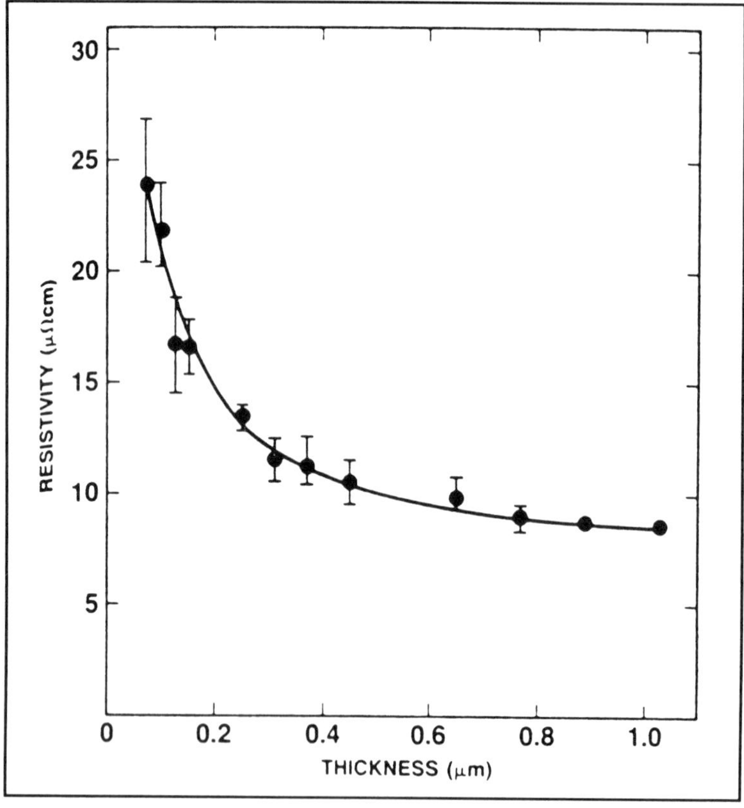

Figure 5.5 Room temperature resistivity versus film thickness. The deposition temperature is 400°C. [Learn et al.[180], reprinted with permission, © 1985, Am. Inst. of Physics].

accounting for an increasing resistivity for decreasing filmthickness. Of further interest is that ρ_o depends on the deposition temperature (14 $\mu\Omega$cm at 300°C and 5.3 $\mu\Omega$cm at 600°C). The explanation suggested was that at higher deposition temperatures less defects are generated in the grains.

Metz et al.[181] deposited 1200Å selective tungsten at 300°C using the H_2/WF_6 chemistry. They found a high residual resistivity (= resistivity at 0°K) of 12$\mu\Omega$cm. Auger analysis showed 0.9 at% oxygen incorporation in the film which could possibly account for the high resistivity. Another reason for the high residual resistivity in their samples could be grain boundary scattering since the grain size was approximately 300Å which is well below the mean free path length for electrons in tungsten of 400Å.

Metz and Beam[182] found a resistivity of 9.4$\mu\Omega$cm for 7500Å tungsten deposited at 600°C using a H_2/SiH_4/WF_6 mixture. The film contained 4% Si.

Van der Jeugd et al.[183] came basically to identical results for the H_2/SiH_4/WF_6 chemistry. At 500°C they found a resistivity of 9 $\mu\Omega$cm. The same value they found for the H_2/WF_6 chemistry. The temperature dependence of the resistivity was found to be 0.03$\mu\Omega$cm/K. This implies that at typical IC operating temperatures (say 100°C), the resistivity will be increased another 3$\mu\Omega$cm.

5.3.6 Determination of the Resistivity

The most convenient way to determine the resistivity routinely is by the four point probe method as described by Valdes[254]. The resistivity can be calculated from the sheet resistance by:

$$\rho = R_\square\, t \qquad (5.13)$$

where $R_\square$ is the sheet resistance and t is the thickness. Unfortunately, for rough films this method will give less accurate results since in this case a certain part of the rough top layer will not contribute to the conductivity in the same way as a specular film would. As a result the bulk resistivity determined according to equation 5.13 will give an overestimation of the true values.

CHAPTER VI

THE CHEMISTRY OF CVD-W AND PROPERTIES OF TUNGSTEN

In this chapter we want to highlight some details of the chemistry which is involved with the CVD of tungsten. Several chemical routes available are available to come to the desired material. The quality of a CVD process is strongly dependent on a proper choice of the deposition chemistry and various deposition parameters. In many cases the allowable deposition temperature is a very important boundary condition and can have a major influence on what chemistry will be chosen. A good knowledge of the chemistry will be of great help in matters like:

- Preventing gas phase reactions (reactivity of the reactants)

- Determining if the chemistry of choice possibly can exhibit a selective nature

- Determining what vacuum requirements are needed (sensitivity of the reaction towards oxygen or moisture).

Further help can sometimes be obtained from thermodynamic considerations (see for example the discussions in section 3.4.2 and section 9.4).

In addition, some physical and chemical properties of tungsten will be highlighted such as wet etching and metallurgy.

6.1 CVD TUNGSTEN SOURCE MATERIAL

At least three different tungsten sources have been reported to give CVD tungsten films: WF_6, WCl_6 and $W(CO)_6$. Some properties of these sources are listed in table 6.1.

Table 6.1

Some physical properties of tungsten sources			
Source	state*	vapor pressure	boil. p.
WF_6	liquid	880 Torr(21°C)	17°C
WCl_6	solid	0.7-7 Torr (150-200°C)	-
$W(CO)_6$	solid	10-50 mTorr (30°C)	-

* at room temperature

From this we see that WF_6 is the most convenient source in terms of vapor pressure and physical state. Generally, solid sources are much more difficult to deliver to the reactor in a reproducible way. The halides are very stable and decompose only at high temperatures whereas the carbonyl compound will decompose readily at temperatures above 200°C. In the following section we will discuss tungsten deposition results obtained with each source.

$W(CO)_6$: [Diem et al.[169c], Kaplan et al.[170], Vogt[171]]. This tungsten source has the clear advantage in that silicon erosion will not occur as will happen in the other two cases (see below). The tungsten is deposited at substrate temperatures of greater than 200°C according to the decomposition reaction:

$$W(CO)_6 \ \text{-------}> \ W + 6CO \tag{6.1}$$

The quality of the tungsten film is poor in terms of purity. The carbon and oxygen content are very high, typically about 20 at.% each. This can lead to a high resistivity: 50-500 $\mu\Omega$cm. Also, the combination of H_2 and $W(CO)_6$ has been reported [Vogt[171]] but the presence of hydrogen had no positive effect on film properties.

WCl_6: [Melliar-Smith[168], Hashimoto et al.[169], Hårsta et al.[169a]]. An important reason for investigating WCl_6 as the tungsten source instead of WF_6 is the Si encroachment problem in selective tungsten. (Although this can be solved by using SiH_4 as the reducer with WF_6, see chapter III). Chlorine etches silicon much slower than fluorine does (the boiling point of $SiCl_4$ = 58°C and SiF_4 = -97°C). Therefore, one would expect the encroachment problems to be less severe for the chlorine chemistry than for fluorine chemistry. Experimental verification of this hypothesis by Hårsta et. al.[169a] showed, however, a severe silicon encroachment using Ar/WCl_6 or H_2/WCl_6 chemistries for selective depositions. This could be explained by the formation of very porous tungsten. In the case of Ar/WCl_6 the encroachment was worse than for H_2/WCl_6 which might be due to the following complication (which does not occur in the presence of H_2):

$$WCl_6(g) + SiO_2(s) \ \text{-------}> \ WO_2Cl_2(g) + SiCl_4(g) \tag{6.2}$$

followed by:

$$2WO_2Cl_2(g) + 3Si(s) \ \text{-------}> \ 2SiO_2(s) + 2W(s) + SiCl_4(g) \tag{6.3}$$

We see that these reactions will sustain themselves since the SiO_2 formed in reaction 6.3 can initiate reaction 6.2. The films deposited by the hydrogen reduction contained typically 0.05-0.1 at.% chlorine.

Because of practical problems such as the sensitivity of WCl_6 to moisture and air, Shroff[169b] proposed in situ generation of WCl_6 by the reaction between the metal and chlorine at about 900°C in a separate chamber. The tungsten chloride is then transported to the deposition chamber where reduction with hydrogen leads to tungsten deposition.

WF$_6$: The majority of work which has been reported on CVD-W uses WF$_6$. This source material has already been used for years as the tungsten source for CVD-WSi$_2$ [see chapter IX]. The compound is fabricated by the reaction between tungsten and fluorine. After several purification steps a very pure product can be obtained (99.999%). Although the gas is dangerous, handling is fairly easy provided that good gas handling practices are applied. Unfortunately, the price of WF$_6$ is high which is the main reason for the high costs of the blanket tungsten fill process. Depending on process conditions, WF$_6$ costs can be 50% of the total costs of the blanket tungsten deposition. Fortunately, CVD equipment is improving in terms of WF$_6$ utilization. Some pertinent physical properties of WF$_6$ are listed in table 6.2.

Table 6.2

Physical properties of WF$_6$

Boiling point	17.1°C
Liquid density	3.44 gr/cm^3
Vapor pressure	887 Torr (21.1°C)
Freezing point	2.0°C

Apart from the reactions mentioned in chapters II and III some other interesting reactions are [Cotton et al.[268]]:

$$W + 5WF_6 \text{ -------> } 6WF_5 \qquad (6.4)$$

This reaction can occur between ca. 550°C and 750°C and the liberated WF$_5$ can be quenched at the cold wall. Above 50°C, WF$_5$ will disproportionate into WF$_4$ and WF$_6$ according to:

$$2WF_5 \text{ --------> } WF_4 + WF_6 \qquad (6.5)$$

WF$_5$ can form a tetramer; (WF$_5$)$_4$ which has a yellow color. WF$_4$ has a red brown color and is reasonably stable and non-volatile.

Extreme care must be taken when using WF_6. Not only with regard to safety but also with regard to prolonged (residual) memory effects which can exist once the reactor or gas lines are contaminated with oxygen or water vapors. For instance, WF_6 can react with H_2O:

$$WF_6 + 3H_2O \quad \text{-------} > \quad WO_3 + 6HF \qquad (6.6)$$

The problem now is that WO_3 can react with WF_6:

$$WO_3(s) + 2WF_6(g) \quad \text{-------} > \quad 3WOF_4(s,g) \qquad (6.7)$$

WOF_4 is believed to inhibit tungsten growth. It has, at $21^\circ C$, a vapor pressure of about 1 Torr! In the pure form it is white but in practice it is usually contaminated with other tungsten oxides which can give it different colors (blue-yellow).

6.2 EXPERIMENTAL DEPOSITION RATE RELATIONS OBTAINED FOR THE H_2/WF_6 CHEMISTRY

It is a common practice to use the thickness of the deposited film divided by the deposition time to represent the deposition rate. This is in principle incorrect. The reaction rate in the heterogeneous kinetic rate theory should be expressed in terms of moles/sec cm^2 or in similar units. Only when it is verified that there are no density and/or compositional changes in the experimental window of interest can one exchange the reaction rate in moles/sec cm^2 by the deposition rate in nm/sec. The determination of the deposition rate, however, needs some further clarification.

The deposition rate can be defined as:

$$\text{Deposition rate} = [\partial(\text{thickness})]/[\partial(\text{time})] \qquad (6.8)$$

and this is in general not identical with:

114

$$\text{Deposition rate} = \text{thickness/deposition time} \qquad (6.9)$$

because there can be a positive (blanket tungsten) or negative (sometimes the case with selective tungsten) nucleation time. Therefore, we need to use expression 6.8 where the thickness is determined as a function of the deposition time. The slope obtained in a plot of thickness versus deposition time will give the correct growth rate which can then be used for further evaluation and kinetic interpretation(see also figure 6.1). As pointed out in chapter VII, the temperature of the wafer may drift during the deposition because of a change in emissivity. In that case the plot of thickness against time is not necessarily linear which makes the determination of the growth rate difficult.

There appears to be a large difference in growth rates when different studies are compared. In table 6.3, four different studies are compared at 285°C and $P_{H2} = 750$ mTorr.

Table 6.3

Comparison of the growth rate of several studies.

Study	Rate[*]	Method
Pauleau [1985]	71	(Stylus)
Broadbent [1984]	34	(B-back scatter)
McConica [1986]	58	(Stylus)
Cheung [1972]	43	(Weight gain)

---

values of Broadbent, McConica and Cheung are calculated from their kinetic expressions; between parenthesis thickness measurement method;
* rate in Å/min, 280°C, $P_{H2} = 750$ mTorr.

From data in the literature, the following kinetic expressions can be obtained:

Broadbent et al.[44]:

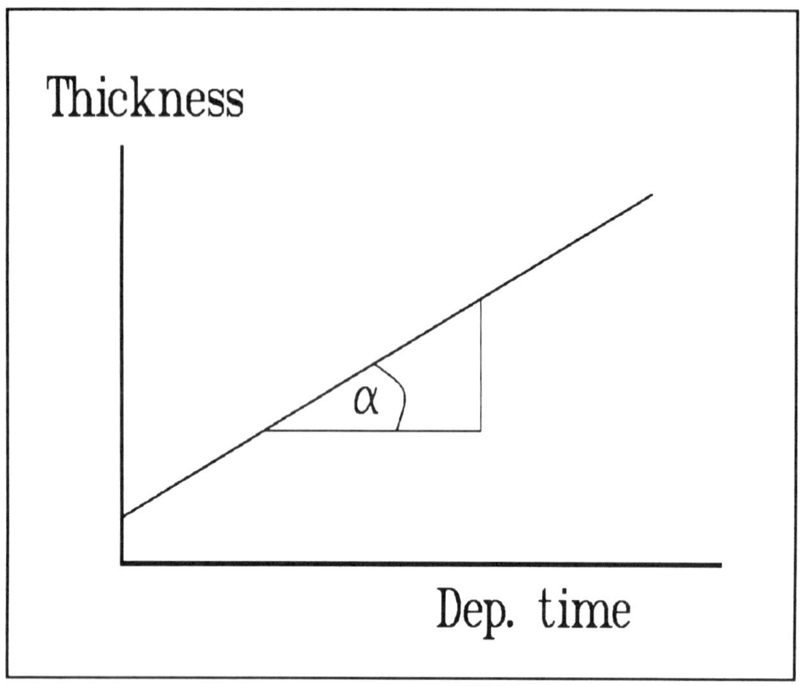

Figure 6.1. Determination of the true growth rate from the slope of film thickness versus deposition time.

$$\text{Dep. Rate} = 7.2\text{x}10^6 \; e^{-8130/T} \; [P_{H2}]^{1/2} \tag{6.10}$$

Cheung[47]:

$$\text{Dep. Rate} = 7.8\text{x}10^6 \; e^{-8040/T} \; [P_{H2}]^{1/2} \tag{6.11}$$

and McConica et al.[46]:

$$\text{Dep. Rate} = 4.1\text{x}10^7 \; e^{-8800/T} \; [P_{H2}]^{1/2} \tag{6.12}$$

All the rates are in Å/min and the hydrogen pressures in Pascal (1 Torr = 133 Pa). Cheung showed that his equation, which is very close to that obtained by Broadbent, holds up to hydrogen pressures of 300 Torr!

The data of McConica et al. was carefully determined in a cold wall

116

reactor whereas Broadbent et al. and Cheung et al. used hot wall systems. To calibrate the wafer temperature McConica et al. used thermocouples attached to the wafer surface. It has been shown in the literature [Blair et al.[172]], however, that in the 600 K range, temperature differences between the thermocouple reading and the actual temperature of the wafer of ca. 10 degrees may exist. Such an error will indeed give a higher pre-exponential factor for the cold wall reactor data. The difference in deposition rate between cold and hot wall reactors, however, seems to be too large to be explained solely by this temperature measurement error.

Another reason for the observed discrepancy may be the thickness measurement of the as-deposited film. The stylus technique has the disadvantage that rough films cause an overestimation of the actual film thickness (see figure 6.2). This again leads to an overestimation of the pre-exponential factor. The weight gain method needs a density of the film. The density of poly-crystalline films is normally less than that of mono-crystalline material. Therefore, a density determination is required before the weight gain method can be used.

6.3 SOME PROPERTIES OF TUNGSTEN

Oxidation Behavior: Tungsten is susceptible to oxidation by oxygen at temperatures above 300°C. The reaction is not self-limiting and is enhanced because its forms WO_3 which is volatile at higher temperatures. Therefore, when tungsten is used as an interconnect material, precaution is needed to prevent oxidation; especially during subsequent dielectric depositions.

Thermodynamically, it is expected that tungsten in contact with SiO_2 will be stable. This has been confirmed experimentally by Krusin-Elbaum et al.[202]. In their study, the integrity of (sputtered) tungsten on gate oxide was investigated. It was shown that the tungsten-oxide interface remained chemically and mechanically stable even after an anneal in He for 30 minutes at 1000°C (see also chapter VIII). Therefore, once tungsten is passivated with an SiO_2 layer there appears to be no limitation to the post processing thermal budget.

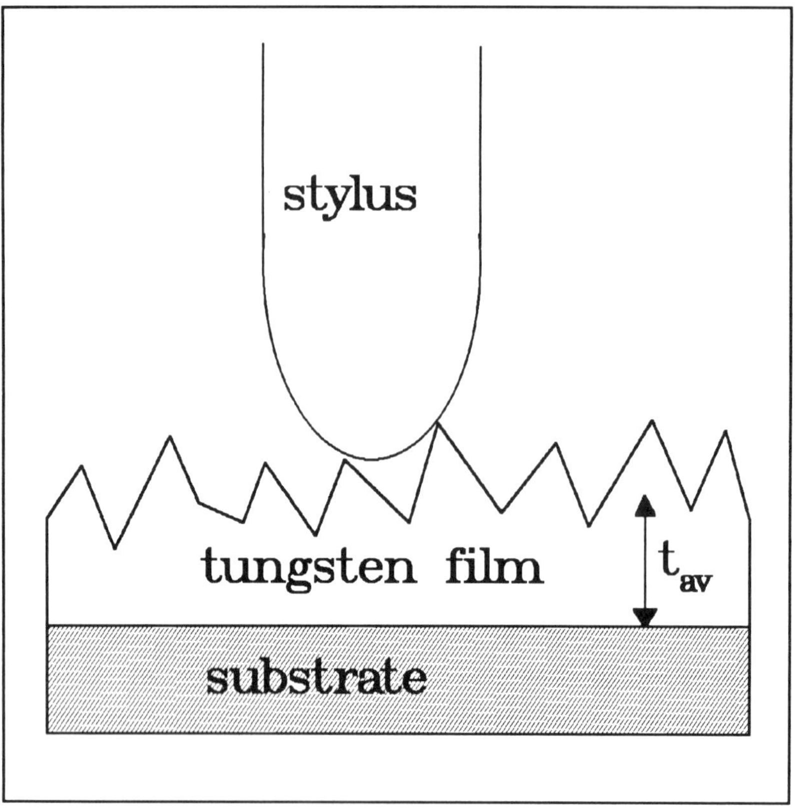

Figure 6.2 Overestimation of the actual film thickness (t_{av}) in the case of rough films and using the stylus technique.

Tungsten can be oxidized at room temperature by (violent) reaction with fluorine under formation of WF_6. This is in fact the chemical route for the synthesis of WF_6 from raw tungsten.

Wet Chemistry: For sub-micron interconnect systems wet etching of tungsten is not a viable way to pattern tungsten lines. However, to clean reactor parts, or to reclaim or repair (8") precious wafers, a good wet etch chemistry is of importance.

Tungsten dissolves very slowly in inorganic acids. The following are examples of wet tungsten etch solutions with acceptable etch rates:

- Hydrogen peroxide (H_2O_2). The reaction is:

$$W + 3H_2O_2 + 2OH^- \longrightarrow WO_4^{2-} + 4H_2O \qquad (6.13)$$

The reaction can be vigorously accelerated by the addition of NH_4OH, however, extreme precautions should be taken since the reaction is exothermic. Therefore, appropriate cooling of the reaction vessel is necessary when using this solution. Etch rates depend on exact conditions but can be as fast as several microns per hour.

- $K_3Fe(CN)_6$. The underlying reaction is this case is:

$$W + 6Fe(CN)_6^{3-} + 8OH^- \longrightarrow WO_4^{2-} + 6Fe(CN)_6^{4-} + 4H_2O \qquad (6.14)$$

Alternative (electrochemical) ways can be found in Kern et al.[258] and references listed in their work.

Pertinent Metallurgical Compounds: In most tungsten IC applications, tungsten will be in contact with metals like aluminum and gold (bonding). In order to have a stable interface the intermetallic compound formation needs to be investigated. Thermodynamically, the following Al-W compounds are possible: WAl_{12}, WAl_5 and WAl_4[de Boer et al.[257]]. At 650°C tungsten up to 1.5 wt% (0.25 at%) can dissolve in Al [Hansen[256]]. Only a few studies have characterized the interaction of CVD-W and aluminum. Thomas et al.[110] found that tungsten obtained via the H_2/WF_6 chemistry reacts with (undoped) evaporized aluminum to form WAl_{12} (from X-ray spectra) above 550°C. Körner et al.[259] saw no interaction between AlSiCu (composition unknown) and etched back tungsten plugs (450°C in 15 min forming gas). Thus, it appears that no additional barrier between CVD-W and aluminum is necessary.

In contrast, there are no known compounds between gold and tungsten. This might have some repercussions for the bonding of gold wires to tungsten bond pads.

It is interesting that tungsten can form compounds with major dopants such as [de Boer et al.[257]):

- For boron: WB_4, W_2B_5, WB and W_2B.
- For phosphor: WP_2, WP and W_3P.
- For arsine: WAs_2, W_2As_3 and W_4As_5.

Tungsten will not react with nitrogen up to temperatures of 1400°C [Hansen[256]). In contrast, the reaction between tungsten and ammonia starts at 140°C and forms a W_2N compound. Broadbent et al.[255], annealed tungsten in NH_3 and found a stabilizing effect on the tungsten-silicon interface with this treatment. A similar effect was found by Deneuville et al.[255].

Also well known of course are the tungsten silicides WSi_2 and W_5Si_3 [see chapter IX] which can be formed by the reaction of CVD-W and silicon. Most of the time this reaction is unwanted because of junction penetration issues. In the case that the tungsten is "stuffed" with contaminants such as oxygen [Thomas et al.[110]] or nitrogen [Smith et al.[260]], this reaction can be blocked up to temperatures of 950°C.

Finally, in table 6.4 some relevant properties of tungsten and silicon are listed. An important fact is that meta-stable β-W has quite different properties than α-W. (Refer to chapter II about the occurrence of β-W in CVD-W films).

6.4 CONTAMINATION ISSUES IN CVD-W

The deposition of tungsten by CVD is essentially a catalytic heterogeneous reaction. The tungsten surface acts as the catalyst to activate either the H_2 or the SiH_4 molecules depending on what chemistry is in use. It is well known from heterogeneous catalysis that extremely low concentrations of surface active contaminants can deactivate the surface and block or slow down the reaction rate. However, it is also possible that certain active molecules can accelerate the deposition once they become adsorbed to the tungsten surface.

Numerous contamination sources can exists in a reactor if not

120

properly designed or maintained. Here we will discuss some possibilities. The first contamination source can be WF_6 itself. Much work has been done in the early stages of commercialization of CVD-W to purify the tungsten source as much as possible. A purity of 99.999% can routinely be delivered. One report mentioned the effect of chromyl fluoride (CrO_2F_2) on selective tungsten [Aitchison et al.[210]]. It was found that 1.5% CrO_2F_2 in WF_6 was able to completely suppress tungsten deposition. Also, films with a high density of surface defects were observed.

Table 6.4

Some physical properties of tungsten and silicon

Property	Si	W	
Density (g/cm^3, 25°C)	2.32		19.32
Young Modulus	11.26 10^{11}		34 10^{11}
Lin. expans. coeff.(K^{-1})	2.5 10^{-6}		4.6 10^{-6}*
	4.0 10^{-6}		4.8 10^{-6}@
Thermal cond. (W/m°K)	65		141
Lattice const. (Å)		α:	3.16517Å
			(bcc, A2)
			Z=2
		β:	5.048Å (.., A15)
			Z=8
Resistivity ($\mu\Omega$ cm)		α:	4.82 (0°C)
			5.33(22°C)
		β:	300-1000
Optical constants:(436nm)		n	3.31
		k	2.47

*=at 20°C; @=at 407°C; Z=tungsten atoms in cell

Another problem is that WF_6 can become contaminated in the reactor by reaction with stainless steel parts or elastomers. George et al.[207]

and Bell et al.[208] show that WF_6 will react with many types of stainless steel; 302, 304, 316L or 318ELS. At room temperature the main product was WOF_4. It was shown that surface passivation (of the gas delivery system for example) using fluorine can be very effective in preventing this. Both WOF_4 and CrO_2F_2 can be formed by the reaction of WF_6 and CrO_3 (which forms the surface of some steels) [Hogle et al.[209]]:

$$CrO_3 \ + \ WF_6 \ \text{-------->} \ WOF_4 \ + \ CrO_2F_2 \tag{6.15}$$

Hogle et al.[209], showed that WF_6 will interact with elastomers such as Vespel and Kel-F. The products, detected using infrared absorption spectroscopy, were CO_2 and CF_4.

Hirase et al.[211], studied the effect of impurities and byproducts on selective tungsten deposition (H_2/WF_6). Oxygen, HF, pump oil and SiF_4 were intentionally injected. Oxygen at the 0.1% concentration reduced the deposition rate. Up to 0.5% concentration no oxygen was found in the tungsten film. Above 0.5%, WOF_4 was detected in the film and the resistivity increased. Added HF decreased the deposition rate at up to 40% concentration, with no effect on selectivity. Oil and other organic molecules (like CH_4 and CO_2) strongly depressed the deposition rate but had no effect on selectivity. SiF_4 up to 10% had no effect on deposition rate or selectivity.

CHAPTER VII

THE DEPOSITION EQUIPMENT

7.1 HOT WALL REACTORS

Many of the first papers which discussed the use of (selective) CVD
of tungsten for IC applications used conventional hot wall tube CVD
reactors [Broadbent et al.[44], Pauleau et al.[45], Cheung[47]]. This type of reactor
was and still is the workhorse in IC fabs. Excellent films such as TEOS
based oxides, thermal silicon-nitride and poly-silicon can be grown in such
equipment. Hot wall tube reactors are suitable for these films because such
materials stick very well to quartz tubes and are quite transparent to IR
radiation of the heating elements. Thus neither particle nor temperature
control is a problem. One other major advantage is that high throughputs
are typically obtained.

This, however, changes drastically when the deposited film is
tungsten. Some problems observed are:

Tungsten peel: Tungsten does not adhere to quartz and even thin
films tend to peel leading to unacceptable particle counts. Although this
could potentially be solved by first depositing an adhesion layer like poly-
silicon, the high stress values of thick tungsten films will soon cause

problems and can even lead to cracking of the quartz tube. Another solution suggested is to clean the tube in situ using either thermally or plasma activated NF_3 cleans [Huggett[271], Kwakman et al.[271]]. Since a frequent clean will be necessary with potentially long conditioning times, this approach does not appear attractive. Another drawback of the use of NF_3 is its price. Therefore, cleaning costs can represent a substantial part of the total process costs.

Temperature Control: As soon as the walls become coated with tungsten there is a tremendous change in the heat transport mechanism. With quartz walls the most important heat transport route is by radiation. This is no longer true with tungsten coatings present since tungsten is non-transparent to IR radiation. Consequently, a majority of the energy from the heater elements is reflected. This can lead to extremely long warm up times for wafers in the tube.

One other problem encountered with CVD-W, and especially in the case of selective tungsten, is that the deposition temperature is rather low (300-400°C). This low temperature is not easy to control in a tube system and special precautions must be taken (thermocouples, heater elements).

Nevertheless, one substantial advantage of the tube systems is that the tube can be considered more or less an isothermal system. This is very advantageous since now the determination of the real wafer temperature is not a problem. This is, as we will see, in contrast with cold wall systems where the real wafer temperature is very difficult to measure and sometimes difficult to control.

Selectivity: Many of the early studies were focused on selective tungsten (based on WF_6/H_2). A clear disadvantage of the tube systems is that the wafers in the rear will see more reaction products than those in the front. As we have seen in chapter III, the reaction products are a major cause for the loss of selectivity. Indeed, poor selective results are normally seen in such furnaces. Another disadvantage of the hot wall system is that as soon as tungsten coating of the wall occurs, there is a tremendous increase of the reaction by-products partial pressures, again leading to poor selectivities.

In the next section we will describe how cold wall reactors can solve some of these problems and, however, introduce new ones.

7.2 COLD WALL REACTORS

In an ideal cold wall reactor the only hot object in the reactor is the wafer surface. All other parts are well below a temperature where deposition can occur. For the H_2/WF_6 chemistry this temperature may be as low as 130°C [Schmitz[266]] and for the SiH_4/WF_6 chemistry even lower in order to prevent tungsten deposition. Therefore, large temperature gradients can exist in cold wall reactors which creates other difficulties such as temperature non-uniformity across the wafer and thermal diffusion effects (vide infra). In the next sections we will address some of these issues.

7.2.1 Heat Transfer

There are at least three ways to heat a wafer in a cold wall reactor:

- by a hot plate on which the wafer resides or
- by inductive (RF) heating or
- by (indirect) lamp heating.

Since the most important industrial CVD-W reactors essentially use hot plate heating we will direct most of our attention to this type of wafer heating.

Generally speaking, there are at least four different routes for transporting heat from one body (the hot plate) to another (the wafer):

- radiation
- gas conduction (diffusion)
- solid-solid contact
- free convection (density differences).

The heat transport by physical contact between the wafer and the hot plate is very marginal as is the case for free convection (certainly at low pressures). Therefore, we will concentrate on the two main pathways, namely radiation and gas conduction (see also figure 7.1).

Heat transport by radiation is described by the Stefan-Boltzmann equation (for two parallel planes):

$$E_r = \frac{\sigma\, \epsilon_1\, \epsilon_2}{\epsilon_1 + \epsilon_2 - \epsilon_1\epsilon_2} (T_1^4 - T_2^4) \qquad (7.1)$$

where E_r is the amount of energy loss by the hot surface in Watts/cm^2, T_1 the temperature of the hot surface, T_2 the temperature of the cold surface, ϵ_1 and ϵ_2 the emissivities of the surfaces and σ is the Stefan-Boltzmann constant ($5.67\ 10^{-12}$ Watt cm^{-2} K^{-4}). We see that the radiation for a given temperature difference T_1-T_2 depends strongly on the emissivities of the participating bodies (see table 7.1).

Table 7.1

Dependence of E_r on emissivity

ϵ_1	ϵ_2	$\epsilon_1\epsilon_2/(\epsilon_1+\epsilon_2-\epsilon_1\epsilon_2)$
0.1	0.1	0.05
0.1	1.0	0.1
0.6	0.6	0.43
1.0	1.0	1

For metallic surfaces (emissivity close to 0.1), heat transfer can be as low as 5% compared to that of black bodies ($\epsilon=1$). We also see that for materials such as silicon and WSi$_x$ (ϵ is about 0.6), heat transfer can be quite efficient and the radiation is about 50% of that of a black body (emissivity 1.0).

The description of heat transfer by gas conduction is less straightforward, especially over a large pressure range. One possible

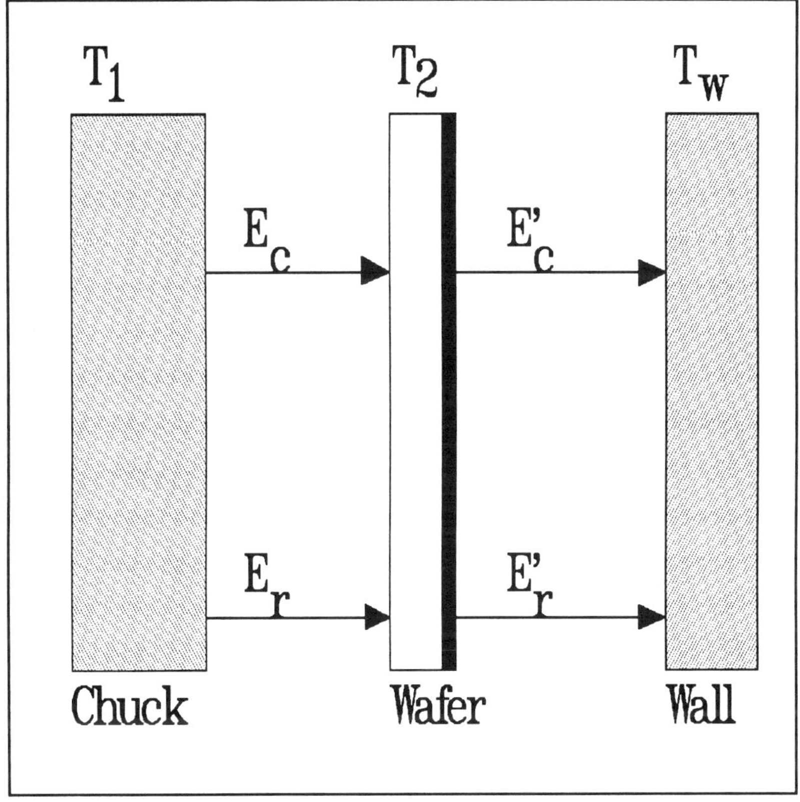

Figure 7.1. Schematic representation of the main heat transfer routes in a cold wall reactor with hot plate heating (see text for details).

approach is that developed by Smoluchowski's temperature discontinuity model. The heat loss of a hot plane surface to another parallel cold surface is given by:

$$E_c = \Lambda \, (T_1 - T_2)/(d + 2 \, \beta \, c/P) \qquad (7.2)$$

where E_c is the heat loss of the hot surface in Watts/cm^2, Λ is the mean conductivity in the temperature range T_1-T_2, d is the distance between the two planes, β is a constant which depends on the gas type and the accommodation coefficient [Dushman[261]] and is of the order of 10, c=LxP, L is the mean free path, and P is the pressure.

Two important borderline cases of gas conduction can be distinguished:

a) At low pressure such that $2\beta L$ becomes comparable to d. E_c varies linearly with pressure in this pressure regime. In practice this is between 0 and 10 Torr, ie. the pressure range of LPCVD!

b) At high pressure such that $2\beta L << d$. E_c becomes now virtually independent of pressure.

The situation is illustrated in figure 7.2. Here the heat loss of a wire in different gases is plotted as a function of pressure. Note that the heat loss at lower pressures depends strongly on the pressure but becomes independent of pressure after about 50 Torr. Thus equation (7.2) is at least in a qualitative sense correct.

Table 7.2

Magnitude of the conductance of gas as a function of the pressure.

P mTorr	$100xE_c$ Watt/cm^2	$100xE_r$ Watt/cm^2
0	0	24
50	2.2	24
100	4.4	24
200	8.8	24
500	22	24
1000	43	24
1.10^4	351	24
1.10^5	1229	24
1.10^6	1637	24

For H_2, T_1=673K, T_2=573
β=9, ϵ_1=ϵ_2=0.6, d=0.01, Λ=1.7 10^{-3} W/cm K.
[From ref. 174, with permission from the Materials Research Society].

Let us have a somewhat closer look at what actually happens when the process pressure in the reactor is varied. We focus on the heat transfer between the hot plate and the wafer and consider only heat transfer by radiation and conduction (diffusion). In addition, we assume for convenience that we have the hypothetical case that the temperature of the wafer and the hot plate are simultaneously constant. In table 7.2 we compare the two

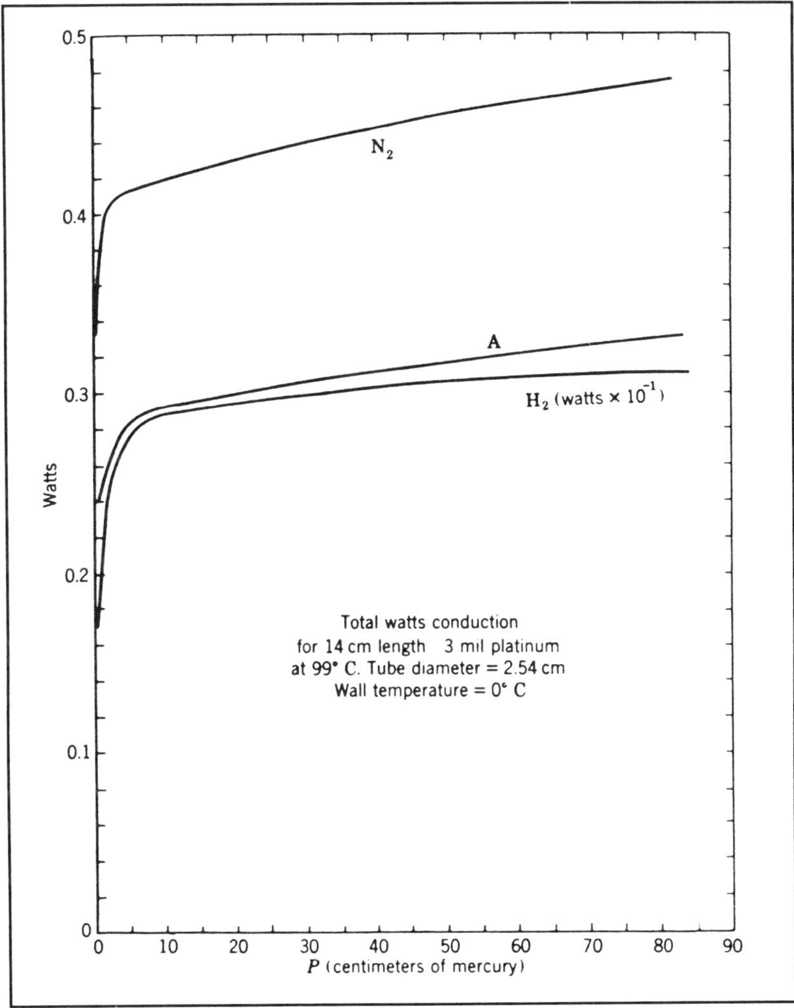

Figure 7.2. Heat loss of a heated wire as a function of pressure. [Dushman[261], reprinted by permission of John Wiley & Sons, Inc., Copyright © 1962].

transport routes at different pressures.

We see that at lower pressures radiation dominates the heat transport. However, at about 500 mTorr the amount of heat transported by either route is almost equal. At 10 Torr radiation accounts for only ca. 10% in the overall transport. This implies that at pressures of 10 Torr and greater, the wafer temperature becomes independent of the emissivities of the chuck and the back side of the wafer. This is nicely illustrated by the data in table 7.3.

Table 7.3

Wafer temperature at base pressure and at 10 Torr in H_2

Chuck coating	Wafer coat.	T_{vacuum} ^oC	T_{10Torr} ^oC
Tungsten	$Si/SiO_2/W$	374	430
Tungsten	$W/SiO_2/Si/SiO_2/W$	346	433
WSi_x	$Si/SiO_2/W$	418	433
WSi_x	$W/SiO_2/Si/SiO_2/W$	358	432

From reference 174, chuck temperature$=482^oC$

First of all, it is important to realize that the emissivities of tungsten, WSi_x, and Si are approximately 0.2, 0.6 and 0.6 respectively. The front coating of the wafer is in all cases sputtered tungsten. The back side of the wafer is either bare Si or sputtered tungsten. We see that the wafer temperature at base pressure follows exactly the emissivity trend (note no transport by conduction!): the high emissivity chuck coated with WSi_x combined with high emissivity Si gives the highest temperature, whereas the low emissivity combination (tungsten chuck - tungsten back side) gives the lowest temperature.

Now we want to have a closer look at the temperature pressure profile between 0 and 10 Torr as demonstrated in figure 7.3. A remarkable

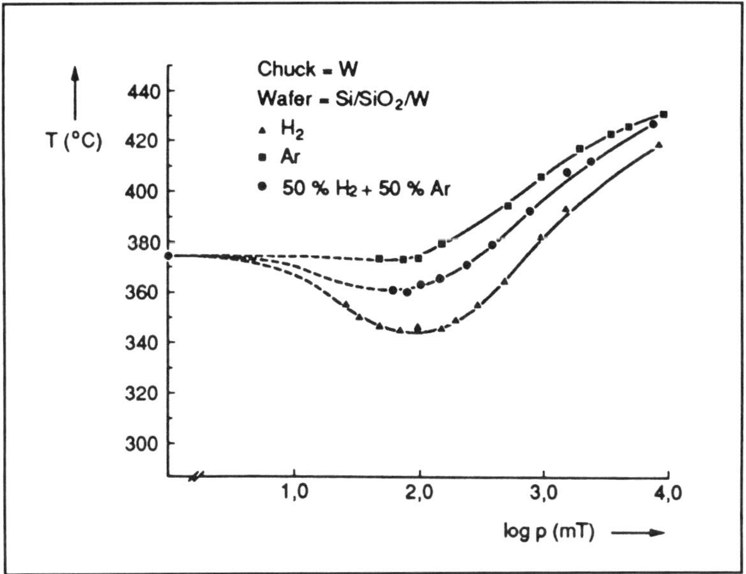

Figure 7.3. Wafer temperature versus pressure. The chuck is coated with tungsten. The backside of the wafer is silicon, the front side is coated with tungsten. [From ref. 174, reprinted with permission].

effect is that in the hydrogen ambient, a drop in wafer temperature occurs on going from base pressure to ca 100 mTorr. Upon further pressure increase the temperature increases and starts to level off at 10 Torr. In order to understand this behavior we have to realize that we now must consider both the heat gain by the wafer coming from the hot plate and the heat loss by the wafer to the cold wall. In fact we need to write equations 7.1 and 7.2 both for the front and the back side of the wafer. In a steady state condition the incoming heat flux and the outgoing heat flux should be balanced. Unfortunately, there is no analytical solution possible for T using these equations. Therefore, we have to use numerical methods. It has been shown [Schmitz et al.[174]] that the use of equations 7.1 and 7.2 indeed predicts a minimum in the wafer temperature-pressure profile.

Some more insight in the temperature-pressure profile can be gained by the following method. Consider the situation at base pressure (i.e. a few mTorr), a chuck temperature of 450^{0}C, a wafer temperature of 350 °C, a chuck-wafer spacing of 0.1 mm, and a wafer-cold wall distance of 10 cm. Now imagine that we perform rapid pressure increases from base pressure to different pressures (say 0, 0.01, 0.02, 0.1, 0.2 and 0.4 Torr). By comparing the incoming and outgoing heat flux of the gas conduction component (remember the radiation is pressure independent), we can predict whether the wafer temperature will increase or decrease compared to the wafer temperature at base pressure (350°C). The results of such a procedure are gathered and shown in table 7.4. Note that the wafer temperature will drop continuously on going from base pressure to about 100 mTorr.

What is the reason that there is there not such a pronounced minimum in the temperature-pressure curve of argon as we see in that of hydrogen (see figure 7.3)? This can be explained by the much lower heat conductivity of argon as compared to hydrogen (1.6×10^{-4} resp. 17×10^{-4} Watt/cm K). Thus the effect of the gas conductance at low pressure in the argon case is about 10 times smaller.

Table 7.4

Effect of sudden pressure increase on wafer temperature				
ΔP(Torr)	L(cm)	Heat in	Heat out	Wafer temp.
0	-	0	0	350
0.01	2.0	5.0	14	<350
0.02	1.0	10.0	23.3	<350
0.10	0.2	49.8	50.0	350
0.20	0.1	99.0	58.3	>350
0.40	0.05	196	63.6	>350

$\beta=5$, $T_{chuck}=450$, $T_{wafer}=350$, $T_{wall}=0^{\circ}$C, $d_{cw}=0.01$ cm, $d_{ww}=5$ cm

Joshi et al.[51], investigated the dependence of the wafer temperature on the pressure for a range up to 55 Torr (see figure 7.4). Note that at pressures above ca. 20 Torr, the difference between the wafer temperature and the hot plate is about 10 degrees. What parameters determine this temperature difference at **high** (i.e. >20 Torr) pressure? To answer this

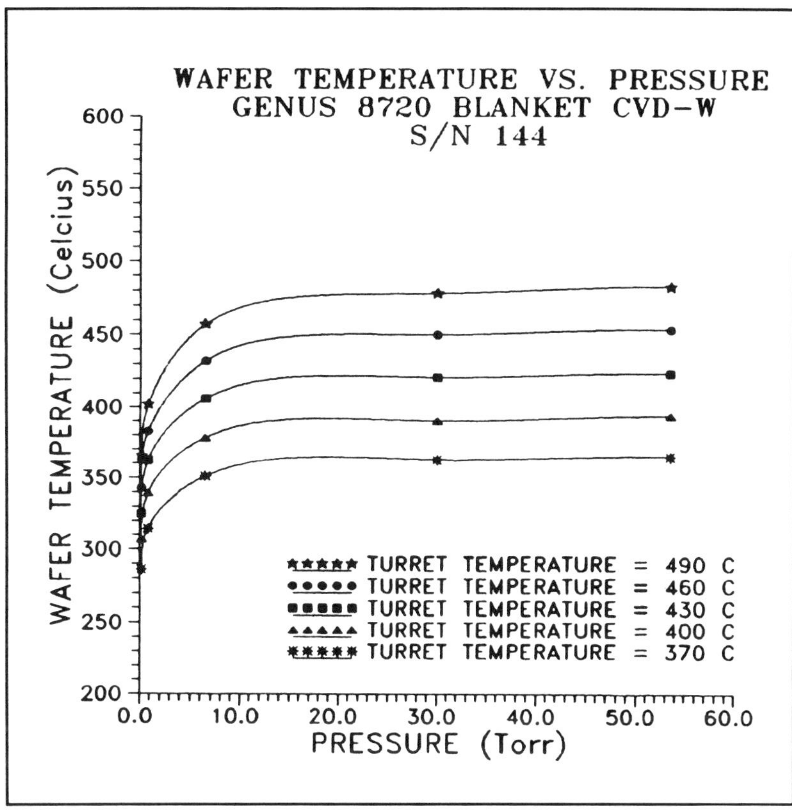

Figure 7.4. Wafer temperature as a function of hydrogen pressure and five hot plate temperatures. [From Joshi et al.[51], reprinted with permission].

question we have to balance the incoming and outgoing heat fluxes of the wafer and solve for the wafer temperature:

$$E_c(\text{chuck-wafer}) + E_r(\text{chuck-wafer}) = E_c(\text{wafer-wall}) + E_r(\text{wafer-wall}) \quad (7.3)$$

Since, as shown above in table 7.2, E_r(chuck-wafer) is small compared to E_c(chuck-wafer) at higher pressures we neglect this contribution in our solution of the equation. After substituting equations 7.1 and 7.2 and solving for the wafer temperature (T_w) we find:

$$T_w = \frac{d_{ww}T_{chuck} + d_{cw}T_{wall} - d_{cw}d_{ww}E_r(\text{wafer-wall})/\Lambda}{d_{cw} + d_{ww}} \qquad (7.4)$$

where d_{ww} is the distance between the wafer and the wall, d_{cw} the distance between the chuck and the wafer, and Λ is the thermal heat conductivity of the gas. Since d_{cw} is usually much smaller than d_{ww} (the wafer-wall distance will be between 1 and 10 cm) the influence of the wall temperature on the wafer temperature is relatively small. Depending on the emissivities of the wafer and the wall, there can be considerable influence of gas type (for instance hydrogen vs. argon) on wafer temperature. Equation 7.4 will be only qualitatively correct since the Smoluchowski equation is not accurate at high pressures. Nevertheless, equation 7.4 gives the correct trend. For the case in figure 7.4 with the chuck at 400°C it predicts about a 2 degree difference between chuck and wafer at higher pressures.

An important conclusion which can be drawn from the discussion above is that in the case of blanket tungsten deposition, the wafer temperature will change during deposition. The adhesion layer will have a different emissivity than the tungsten film. Moreover, the emissivity of the tungsten itself is also a function of film roughness and will therefore change during deposition (roughness increases with thickness). Since the wafer temperature depends on the emissivity via relation 7.2, its temperature will change during deposition. This effect, however, will be more pronounced at lower pressures (<1 Torr) since radiation is predominant there.

Another interesting question that arises is whether there will be a temperature difference between the back and front side of a (silicon) wafer. Consider the situation at 500 mTorr and refer to Table 7.2. The total heat flux towards the wafer is 50×10^{-2} Watt/cm^2. In a steady state this same energy flux will also leave the front side of the wafer. Now we can apply Fouriers law:

$$Q/A = \Lambda \, [dT/dx] \qquad (7.5)$$

$Q/A = 50\times10^{-2}$ Watt/cm^2. Λ for silicon at 573 K is about 50 Watt/cm K. If we assume a linear temperature gradient in the silicon and a wafer thickness of 500 um, we can equate dT/dx to be $dT/5\times10^{-4}$ °C/cm. When we solve for

$_\Delta$T we find its value to be 0.05°C (at 500 mTorr!). This is a very low value and is rather pressure dependent; at 10 Torr $_\Delta$T can become about 2 degrees.

There exists one other heat source in the reactor namely the heat generated by the reaction itself. Consider the (tungsten) reaction with the most negative free energy i.e. the reaction between silane and WF_6. Per mole of tungsten there is about 207 kcal heat generated at 600°K. For a growth rate of 200nm/min, this is equivalent to a heat generation of 3×10^{-2} Watt/cm^2. When we compare this with the heat flux coming off the wafer at 500mTorr of about 50×10^{-2} Watt/cm^2, we conclude that the selective tungsten reaction (based on SiH_4/WF_6 chemistry, see chapter III) could have some (local) effect on wafer temperature. However, one should realize that the active area of the wafer in selective tungsten CVD is mostly in the range of 1-5% of the total wafer area. In the case of hydrogen reduction, the free energy change is one order of magnitude smaller than for the silane case. In this case we cannot expect any influence of the reaction heat on wafer temperature. This is especially true for the high pressure blanket process since the heat loss of the wafer due to radiation and conduction is much higher than at lower pressures.

In the case of direct wafer heating by lamp heating there is a difference in the way heat arrives at the back side of the wafer. Now the main transfer route of heat to the wafer is that by radiation which is independent of the pressure. This implies that the wafer temperature can depend much more on the process pressure since pressure affects only the heat loss from the front side of the wafer. With the hot plate heating both the incoming and the outgoing fluxes are influenced by pressure which leads partly to cancellation. In addition the determination of the wafer temperature is even more complicated than in case of hot plate heating because the wafer has to be monitored directly (thermocouple against wafer or pyrometer).

7.2.2 Temperature and Thickness Uniformity

In the case of blanket tungsten the thickness uniformity across the wafer is of primary importance (see also chapter II). The temperature distribution across the wafer is a key parameter which influences thickness spread. The effect of temperature on deposition rate can easily be calculated by calculating the deposition rate at several temperatures using equation 2.3 and for $E_a/RT = 8800/T$. Normalized values are given in table 7.5.

Table 7.5

Normalized growth rates ($400^{\circ}C = 1$)

T ($^{\circ}C$)	Growth rate
390	0.82
400	1.00
410	1.21

Roughly speaking we see a 2% variation in growth rate with each degree change in wafer temperature. Therefore, if the uniformity needs to stay within 10% only a $5^{\circ}C$ temperature spread across the wafer is allowed. With a proper design of the hot plate this is easy to accomplish.

7.2.3 Thermal Diffusion

The effect of thermal diffusion or Soret diffusion was originally predicted by Soret for liquid solutions. In 1917 the effect was also predicted for mixtures of gases by Enskog, and by Chapman and Dootson. Basically, thermal diffusion can provoke a separation of the components of a mixture under influence of a thermal gradient. There is some relationship between thermal diffusion and the movements of particles in a thermal gradient, known as thermophoresis. A particle suspended in a fluid subjected to a thermal gradient will exhibit collisions from hot molecules at one side and cold molecules from the other side. Thus there will be a net force of the particle directed to the cold side. Because of the mathematical complexity this subject is treated only briefly in most text books. Unfortunately, as will be shown below, the effect can be very substantial in WF_6/H_2 mixtures. Since a comprehensive treatment is beyond the scope of this book, we will give only a few basic equations and refer to the literature [see the compilation given in References] for the interested reader.

The following simple mathematical description of the thermal diffusion itself is rather straightforward. The mass flux due to thermal diffusion has been described by Bird, Stewart and Lightfoot[160] and Wahl[165]:

$$j_i^T = - D_i^T \bullet d[\ln T]/dx \qquad (7.6)$$

here j_i^T is the mass flux of species i in the mixture, D_i^T is the multi-component thermal coefficient of species i, and the sum of them should be zero:

$$\Sigma\ D_i^T = 0 \qquad (7.7)$$

Therefore, in a two component mixture one component will diffuse to the cold region (the heavier one ie. WF_6) and the other component to the hot zone (the lighter one ie. H_2). This thermal diffusion is of course opposed by ordinary diffusion:

$$j_i = - D_i\ d[C_i(x)]/dx \qquad (7.8)$$

and here j_i is the flux due to the ordinary diffusion and D_i is the multi-component diffusion coefficient of species i. In a steady state the following relation clearly holds:

$$j_i^T + j_i = 0 \qquad (7.9)$$

For a two component mixture A and B and using some assumptions one can derive that the amount of separation of A is:

$$x_{A,T2} - x_{A,T1} = - k_T\ \ln[T_2/T_1] \qquad (7.10)$$

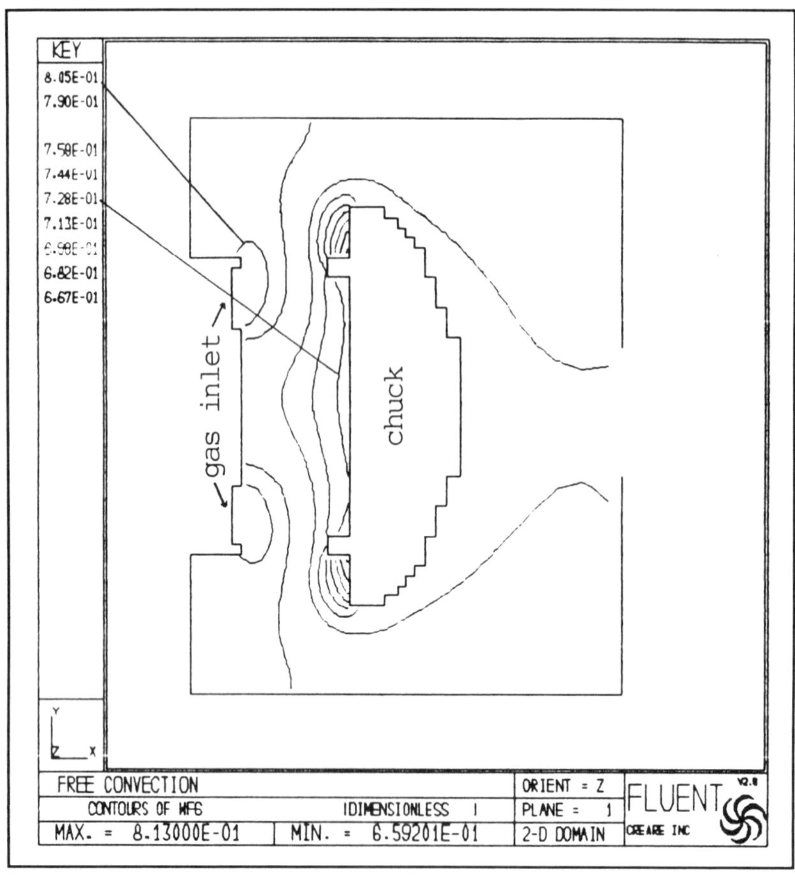

Figure 7.5. Fluent simulation of the WF$_6$ concentration profile at 800 mTorr and at 400°C. [Courtesy of E. Rode, Genus, Inc.].

where $x_{A,T2}$ is the mole fraction of A in the volume at T_2, $x_{A,T1}$ the mole fraction of A in the volume at T_1, and k_T is the thermal diffusion ratio which is proportional to D_A^T/D_{AB}. It can be shown both theoretically and experimentally that k_T is pressure independent. The problem now is to determine k_T. The thermal diffusion ratio can be related to the thermal diffusion factor α by:

$$k_T = \alpha \times x_A \times x_B \qquad (7.11)$$

α is almost independent of the concentration [Bird et al.[160]]. If the

138

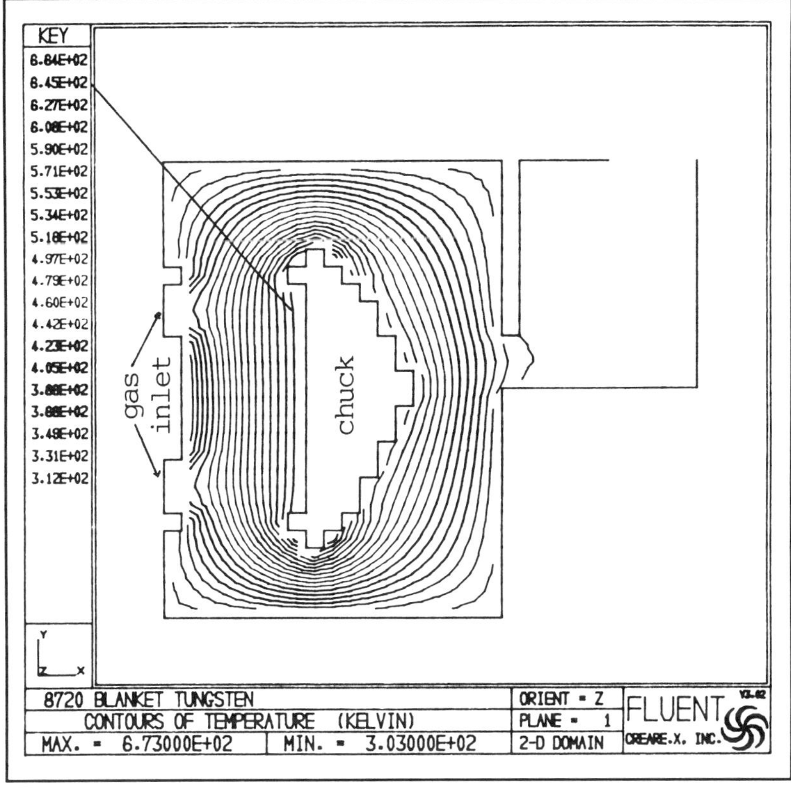

Figure 7.6. Fluent simulation of the temperature distribution at 800 mTorr and at 400°C. [Courtesy of E. Rode, Genus, Inc.].

molecules can be treated as rigid spheres α also becomes independent of the temperature. Wahl[165] showed that under the rigid elastic sphere approximation α depends on the Lennard Jones parameters σ_{AB} σ_{BB} and the molecular masses M_A and M_B. However, in the case of a large and heavy molecule A compared to B (ie. WF_6 compared to H_2) α can be approximated by:

$$\alpha = 0.221 \, [(\sigma_{H2} + \sigma_{WF6})/2\sigma_{H2}]^2 \qquad (7.12)$$

Therefore, for H_2/WF_6, $\alpha = 0.535$ ($\sigma_{H2} = 2.83$ Å and $\sigma_{WF6} = 5.97$ Å). Under typical blanket tungsten deposition conditions $x_{H2} = 0.95$ and $x_{WF6} = 0.05$ thus $k_T = 0.535 \times 0.95 \times 0.05 = 0.03$. Substituting this result in equation 7.10 while using $T_2 = 673$ K and $T_1 = 273$ K gives $x_{WF6,673}$ -

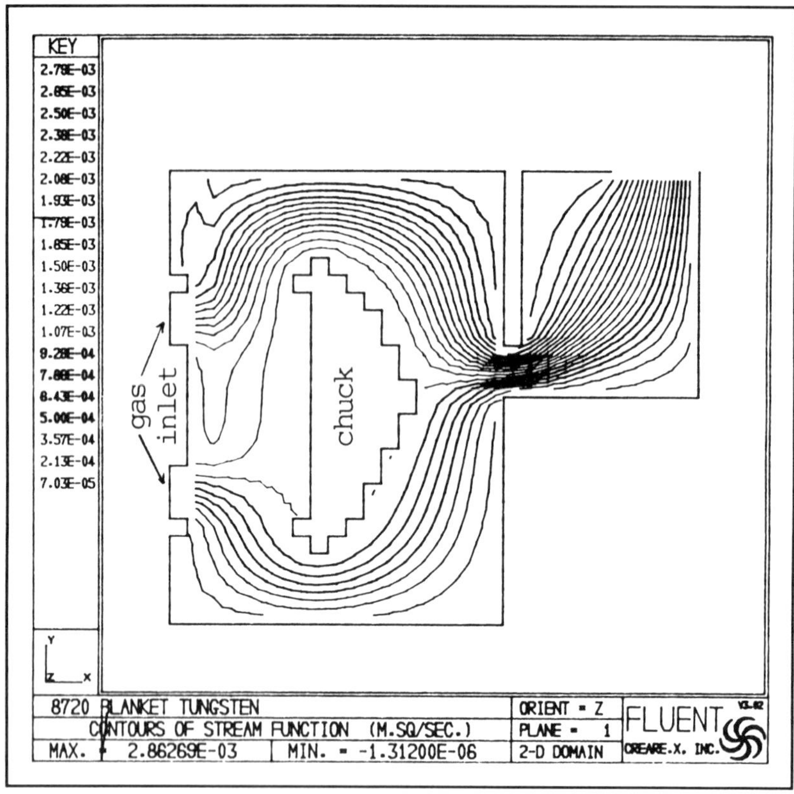

Figure 7.7. Fluent simulation of the streamlines at 800 mTorr and at 400°C. [Courtesy of E. Rode, Genus, Inc.].

$x_{WF6,273} = -0.03$. Thus the mole fraction in the cold area is 0.065 and in the hot area is 0.035. This is roughly in agreement with the results of Kleijn et al.[163] who pointed out, using numerical methods, that the magnitude of separation of reactants can be substantial in the case of H_2/WF_6 mixtures. They found that under the influence of a temperature difference of about 400 °C the difference in mole fraction of WF_6 in the cold and hot region can be as large as a factor of 2! Unfortunately, the thermal diffusion effect tends to deplete the wafer surface of WF_6. This will drive the reactor much earlier to a diffusion controlled regime. As a result the step coverage can degrade substantially under the influence of the thermal diffusion effects as has been found by Hasper et al.[32].

140

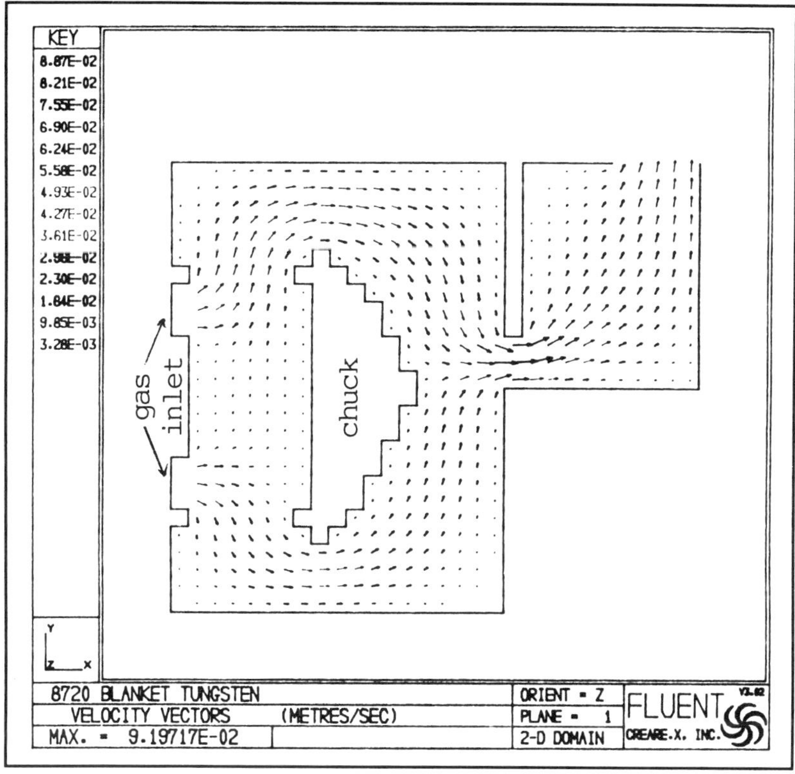

Figure 7.8. Fluent simulation of the gas velocity pattern at 800 mTorr and at 400°C. [Courtesy E. Rode, Genus, Inc.].

7.2.4 Distribution of Temperature, Concentrations and Gas Velocity in a Cold Wall Reactor

It can be very helpful to understand the behavior of a certain reactor if numerical modeling is available. Several studies [Kleijn et al.[163], Ulacia et al.[251]] of cold wall reactors have been made. Here we only want to illustrate the method using the outcome of such numerical calculations. Typical pictures obtained are the following: concentration profile for WF_6 (figure 7.5); temperature distribution in the gas phase (figure 7.6); stream lines (figure 7.7) and gas velocity pattern (figure 7.8). The calculations were done for an 800 mTorr process based on H_2/WF_6 chemistry at 400°C using the Fluent software [Creare[269]]. Thermal diffusion effects are not included.

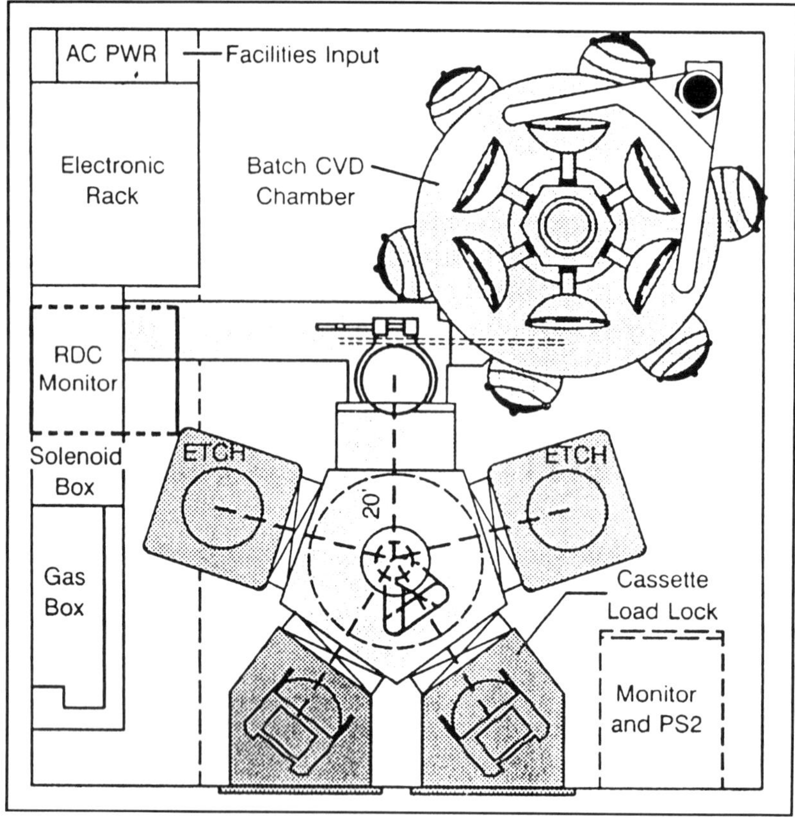

Figure 7.9. Plan view of the Genus 6000 reactor for integrated processing, see text. [Courtesy Genus, Inc. Mountain View].

7.3 INDUSTRIAL REACTORS

7.3.1 Type of equipment

Genus, Inc. pioneered the development of cold wall reactors for high volume production. First (in 1983) for tungsten silicide (see chapter IX) and later followed by both selective and blanket tungsten CVD. Whereas the 8300 and 8400 series were without a loadlock, the second generation

(8700 series) was equipped with a loadlock system. The choice of a cold wall system was forced by the risk of gas phase nucleation in case of the SiH_4/WF_6 chemistry and because of particle considerations. Recently Genus has released their third generation (6000 series, see figure 7.9) reactor which is capable of doing integrated processing (see below).

All other important tungsten deposition reactors (Applied Materials P-5000, Varian 5101, Spectrum 211, Ulvac Era 2000, Watkins Johnson APCVD, Novellus Concept One-W) are of the cold wall type. All reactors are of the single wafer (see figure 7.10) type except the Novellus Concept One-Wand the Watkins Johnson (which are of a continuous nature), and the Genus 8720 and the Genus 6000 reactors (which are of a batch processing nature). These reactors process at sub-atmospheric pressures except the Watkins-Johnson belt type reactor. The advantage of a batch reactor is without a doubt the high wafer throughput under most deposition conditions. Certainly for high volume production the low throughput issue remains a concern of the single wafer approach. The Watkins Johnson reactor is essentially of the single wafer type but has a high wafer throughput because the reactor can continuously process wafers without having load/unload cycles. Although the main emphasis today is blanket

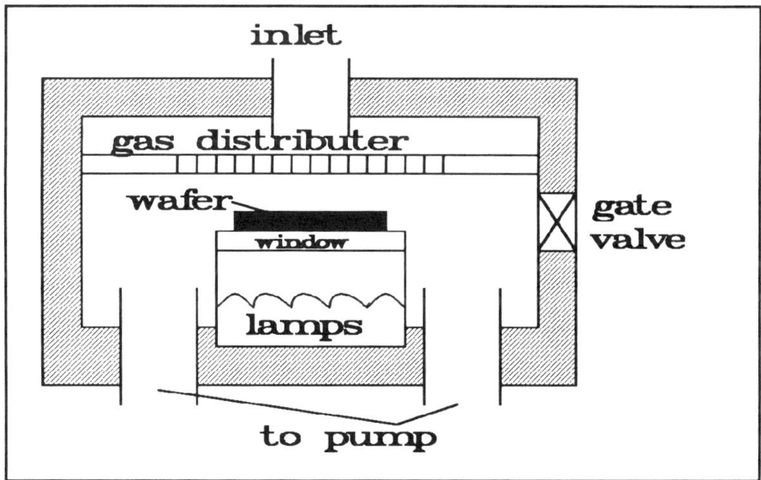

Figure 7.10. Sketch of a single wafer reactor. In this case the wafer is heated by lamps.

tungsten deposition, many systems are capable of performing selective tungsten as well.

7.3.2 Backside deposition

One difficult problem in early blanket tungsten processing was that of backside deposition. In many cases there is no need for an adhesion layer at the back side of the wafer. Especially in the case of an oxide backside, tungsten will flake from the backside if the tungsten deposition is not suppressed in this area. Needless to say, flaking is totally unacceptable from a particle contamination point of view. Several approaches are available today to the backside deposition problem of which three will be discussed.

Genus' clamp method: The backside of the wafer is shielded by a clamp in combination with a backside purge (see figure 7.11). The method is very efficient and is proven in production.

Applied Materials' method: In this approach backside deposition does occur, however, the tungsten at the backside is stripped back using a plasma etch. The etch is done such that the tungsten at the frontside remains untouched.

Novellus' method: Novellus uses vacuum clamping of the wafer center while purging the edge of the wafer with an inert gas (see figure 7.12).

All these methods have their advantages and disadvantages. Nevertheless it can be said that the problem of back side tungsten flaking can be solved by one of these methods and is no longer a serious drawback for blanket tungsten processing.

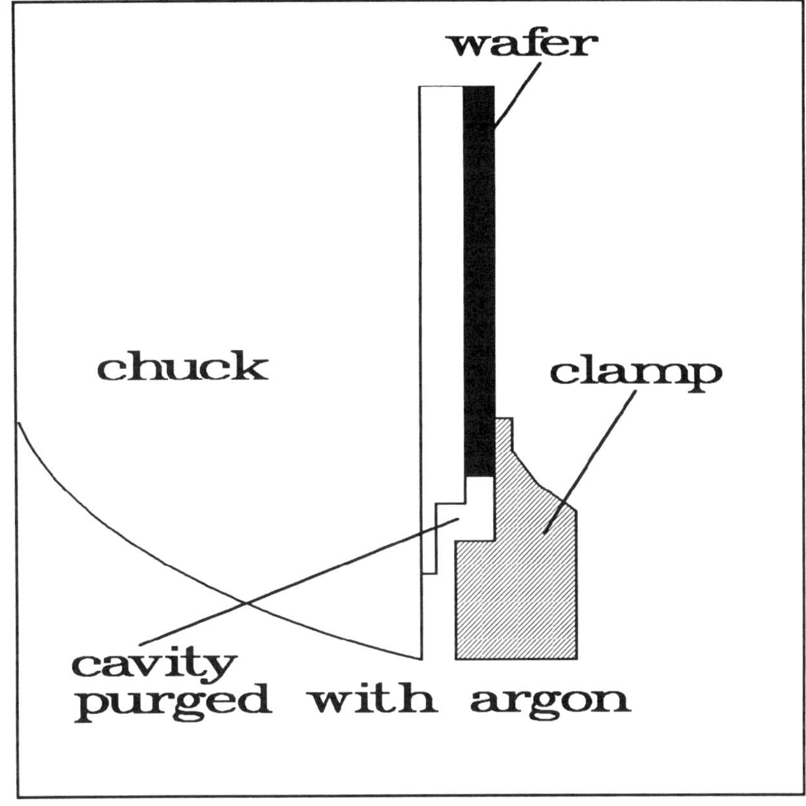

Figure 7.11. Cross-section of chuck with backside deposition prevention using a clamp and a back side inert gas purge (Genus 8720 and 6020 systems).

7.3.3 Particle Contamination

The measures which have to be taken to ensure low particle generation in a tungsten reactor are in general identical to those taken in other (CVD) IC manufacturing equipment. However, the risk of particle capture by the wafer is drastically higher at high pressure than at low (a few milli Torr) pressure. This is because the higher the pressure, the slower the settling speed of the particles [Larrabee et al.[270]]. As an example, a particle of 0.5 um size takes about one minute to drop one meter at 10 Torr! Thus once a particle is generated in the reactor at high pressure it can stay for a

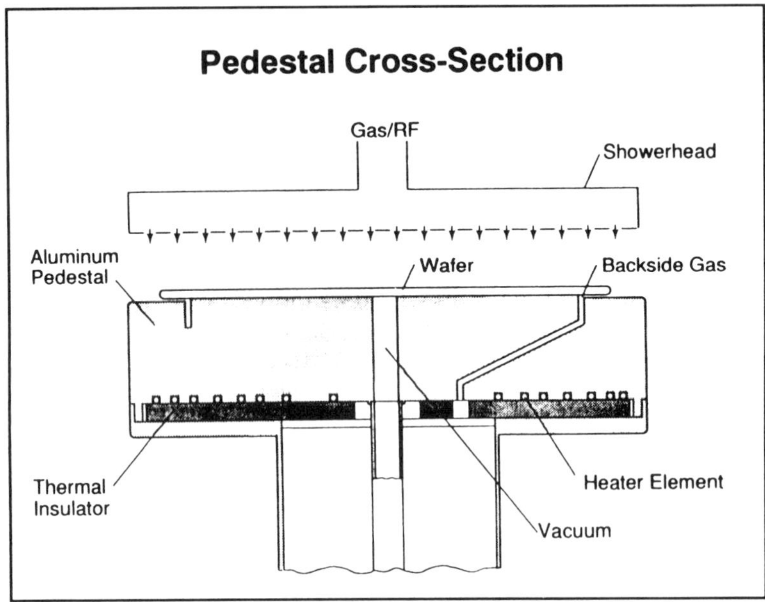

Figure 7.12. Novellus clampless method to prevent back side deposition (Concept One-W system). [Courtesy of E. van de Ven, Novellus].

prolonged time in the gas phase and the chance of becoming trapped by the wafer is therefore larger. Appropriate measures are necesary.

Another complication can be the edge exclusion of the adhesion layer. Some sputter equipment use clamps or fingers to hold the wafer during deposition. In these areas no adhesion layer is present. If tungsten deposition in these areas is not prevented, tungsten peeling can occur causing a major particle source in the reactor or in the processing line.

7.4 FUTURE REACTOR DEVELOPMENTS

The most important development in the near future is the introduction of cluster tools (see figure 7.13) or "integrated processing". At this moment about 20 cluster tools are available for various types of

146

applications [Burggraaf[157]]. Basically a cluster tool is a central handling system with several peripheral process chambers attached such that complete integration of a process module (for instance contact fill) is possible. Each process chamber is supposed to be completely independent in terms of its vacuum and electrical system. It was suggested [VanLeeuwen[158]] that in the year 2010 more than 350 chambers could be linked together. Crucial in this approach is that the central platform be very reliable as should be the case for the integrated process steps.

Among the advantages of cluster tools are:

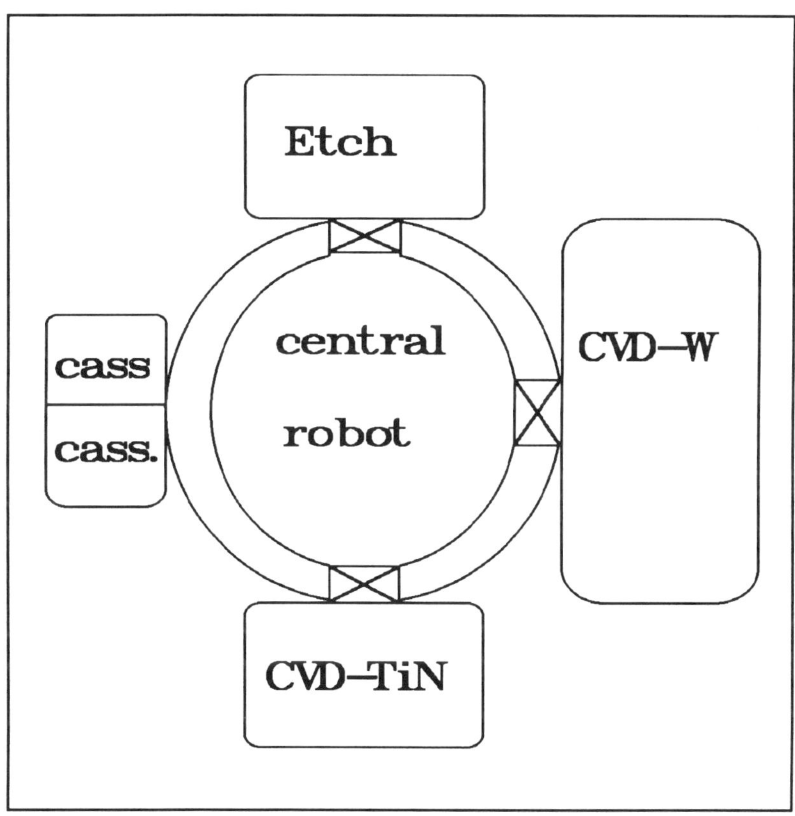

Figure 7.13. Principle of a cluster tool. Via a central loading station several processing units are integrated. The units can be either single wafer or batch type.

- The wafer is not exposed to the clean room ambient between critical process steps . This can result in a more repeatable process. Selective tungsten deposition should benefit greatly from such an approach.

- A larger part of the wafer handling occurs in an ultra clean environment. Therefore there are less transfers between vacuum and atmosphere. This should reduce particle contamination problems and lead to less defects.

- The integrated tool will have a smaller footprint than the stand alone units which provides a better cost performance.

-As the long term process requirements change, the tool can be reconfigured. Thus the life time of very expensive equipment can be extended.

Blanket and selective tungsten plug processes are two of the first candidates for cluster tool integration. In the case of blanket tungsten the necessary process steps are:

1) Clean (sputter etch) of contacts/vias followed by adhesion layer sputter deposition. Instead of a sputtered adhesion layer a CVD adhesion layer like TiN can be applied.
2) Blanket tungsten deposition.
3) Blanket tungsten etchback.
4) An aluminum sputter chamber can also be integrated.

In the case of selective tungsten the process chambers could be:

1) Pre-deposition pretreatment chamber. This chamber can be either a plasma pretreatment chamber [Nowicki et al.[250]], a wet gaseous HF clean [van der Heide et al.[250], Deal et al.[250]] or a methanol vapor exposure [Izumi et al.[250]].
2) Selective tungsten deposition.
3) Depending on the process, a post treatment chamber might be necessary.
4) Here also an aluminum sputter chamber could be included for

the interconnect.

In both cases the wafers are ready to go to the next (interconnect lithography) step.

It is clear that in order to make successful such an extensive process integration, very robust process steps are needed which are very repeatable and work within a wide, stable process window.

CHAPTER VIII

MISCELLANEOUS

In this chapter we will discuss some applications of CVD-W which are presently in a development stage but which might become important in the near future.

In the previous chapters we dealt with tungsten depositions using thermally activated reactions only. Two other techniques will now be discussed namely plasma enhanced CVD of tungsten and the deposition of tungsten by photo activation.

In addition we will briefly discuss some alternatives for the CVD-W plug processes as they are proposed in the literature.

8.1 TUNGSTEN GATES

8.1.1 Problems of Poly-Si Gate Electrodes

The use of n^+ poly-Si as gate electrode material has certain limitations. Among these limitations are:

i) The sheet resistance of this material is typical of the order of 20 $\Omega/\square$, leading to unacceptable delay times in sub-micron devices [Pauleau[203]].

ii) Because of the non-symmetrical work function problems arise in CMOS technology such as punch through in PMOS transistors [Wong and Saraswat[201]]. Although this can be overcome by the use of p^+ poly-Si this leaves the high resistance and introduces p/n junctions in the poly-Si lines where n+ and p+ poly lines merge.

In the following section we will show that tungsten as the metal gate can possibly solve these issues. A very attractive property of refractory metals, such as Mo, W, Ti and Ta, is that they have mid gap workfunctions which is very beneficial in CMOS technology.

8.1.2 Tungsten as The Gate Material

In the discussion below we limit ourselves to pure metal gates although alternatives have been proposed such as a tungsten/poly-silicon stack [Wong and Saraswat[201]]. It is helpful to keep the process flow as depicted in figure 8.1 in mind.

Metal gates have already been proposed studied and used as early as 1960. Issues which were encountered include:

i) Stability of the metal-oxide interface.
ii) The adhesion on gate oxide.
iii) The resistance of the metal gate against oxidation.
iv) Implantation masking capability of the metal gate.
v) Mobile ion contamination in the metal gate which can degrade transistor behavior.

We will briefly discuss the solutions proposed for the issues mentioned above.

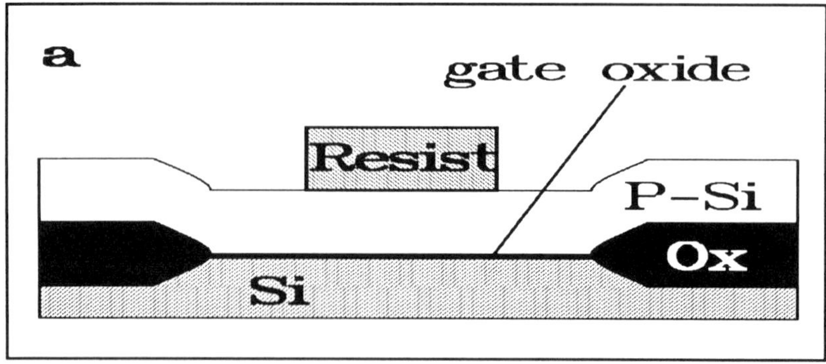

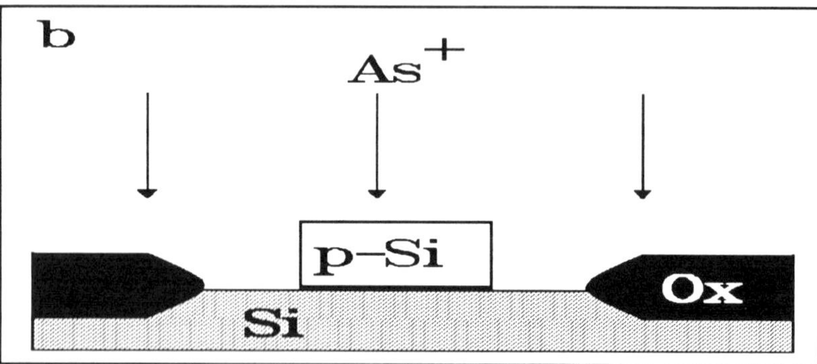

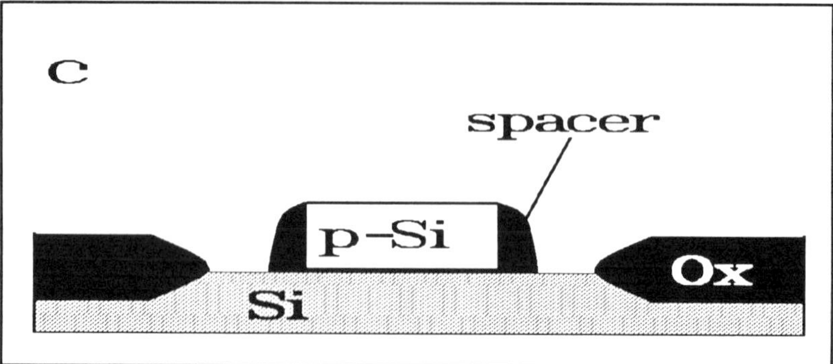

Figure 8.1. Conventional gate process. a) before poly-Si etching, b) implantations, c) oxide regrowth and spacer formation.

152

Miscellaneous.

The stability of the metal-oxide interface: The main problem here is the physical and chemical stability during heat treatment (anneal or oxidation). The metals Ta,Ti, Mo and W have been studied by Kobayashi et al.[200]. Only the W/SiO$_2$ interface was stable for anneals up to 1200°C. Krusin-Elbaum et al.[202] came to a like wise conclusion that only the W/SiO$_2$ interface is acceptable for gate applications in a comparison of Mo/SiO$_2$ and W/SiO$_2$ interfaces. This experimental result is supported by thermodynamic calculations regarding the W/SiO$_2$ interface stability.

Adhesion to oxide: Although sputtered W has excellent adhesion to gate oxide, in the case of CVD-W as the gate electrode material, the adhesion is, for the as-deposited material, rather poor. Kobayashi et al.[200] found that, at anneal temperatures greater than 800°C in either N$_2$ or H$_2$/H$_2$O, excellent adhesion could be obtained. The tungsten was deposited using a H$_2$/WF$_6$ chemistry with the addition of SiH$_4$ in order to diminish the selective nature of the tungsten deposition.

Oxidation resistance: We see in figure 8.1.c that there is an oxidation step after the patterning of the gates. The oxidation is necessary to restore the damage which occurred during etching of the gate oxide at the gate perimeter. Tungsten unfortunately oxidizes readily at temperatures above about 300°C. To overcome this problem, Kobayashi et al.[197] and Iwata et al.[198] developed a wet hydrogen oxidation (WHO) procedure which allows the Si to oxidize while leaving the tungsten unaffected. This method is based on thermodynamic calculations which show that at, for example, 1000°C and a P_{H2O}/P_{H2} ratio of 10^{-5} the equilibrium:

$$Si + 2H_2O \quad <\text{------}> \quad SiO_2 + 2H_2 \qquad (8.1)$$

lies far to the right and

$$W + 3H_2O \quad <\text{-------}> \quad WO_3 + 3H_2 \qquad (8.2)$$

lies far to left. Therefore, under appropriate conditions it is possible to oxidize silicon again such that the oxidation rate of W is very small.

Ion implantation masking: The gate structure as etched in figure 8.1.c is used to self align the implant of the source and drain regions. Poly-Si is a good masking material for the implantation, however, tungsten is not. One approach to solve this is by the deposition of an (PSG) oxide layer atop the tungsten before the gate is patterned. The oxide acts as the implantation mask [Yamamoto et al.[196], Kobayashi et al.[197]].

Mobile ion contamination: Whereas tungsten gates fabricated using sputter deposition need ultra pure targets [Yamamoto et al.[199]], CVD-W (using 99.5% purity WF_6) showed very low (0.01 ppm) mobile ion contamination [Kobayashi et al.[200]].

Two additional advantages of CVD-W over sputtered tungsten are:

i) The excellent step coverage of CVD-W.

ii) The potential incorporation of a little fluorine in the tungsten might improve the gate oxide quality (see Wong et al.[201] and chapter IX).

8.2 SELECTIVE GROWTH ON IMPLANTED OXIDE

Some interesting examples have been described where tungsten has been selectively grown on implanted oxide. The silicon dioxide can be implanted for this purpose with either silicon or tungsten. Both elements can initiate tungsten growth provided their surface concentration is not too low. As will be shown below, a clear advantage of this technique is that (local) interconnects can be made without the need for pattern definition of the metal. In one process step cladding of source, drain and gate can also occur.

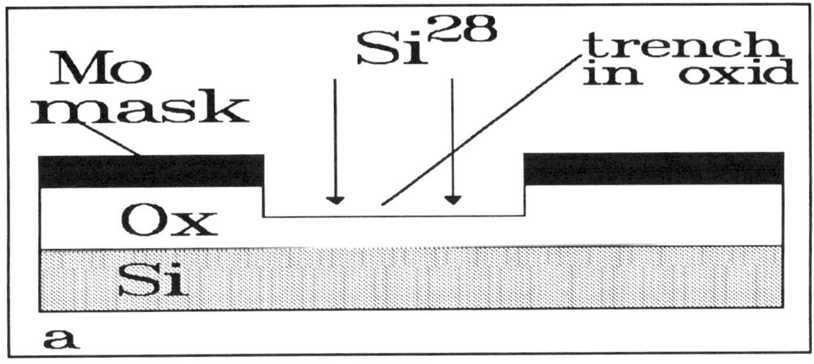

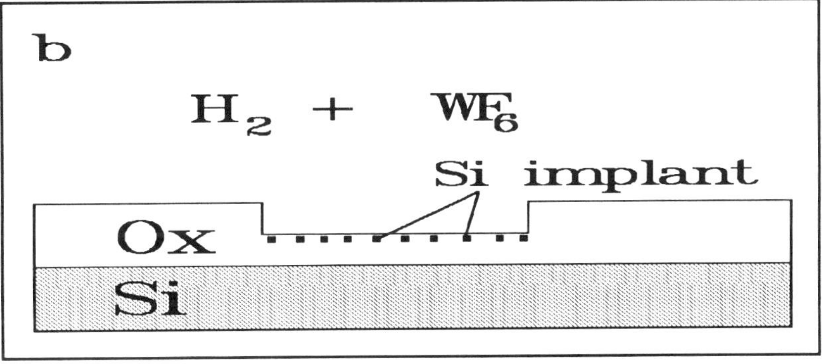

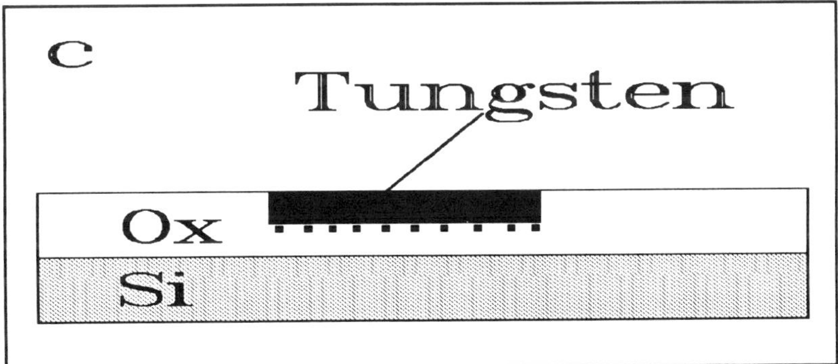

Figure 8.2. Selective tungsten using silicon implantation. a) Mo is used as an implant mask, b) after removal of the Mo mask, c) after selective tungsten deposition.

Miscellaneous.

8.2.1 Tungsten Growth Using Silicon Implants

Hennessy et al.[204] described for the first time tungsten growth on oxide using silicon implants in silicon dioxide. Using molybdenum as a hard mask (see the process flow in figure 8.2), the silicon dose was varied between 1E13 to 2E17 ions/cm^2. The implant energy was 25 eV which gives a projected range of 53 nm with a straggle of 23 nm. After removal of the molybdenum mask, using a wet H_2SO_4/H_2O_2 strip, the SiO_2 is etched back slightly with 1% HF to bring the silicon peak concentration closer to the oxide surface. The following interesting results and conclusions were obtained:

- Below a Si dose of 5E16 at/cm^2 no tungsten growth could be obtained.
- Tungsten film adhesion was good for the doses 5E16 and 1E17 but the 2E17 samples did not pass adhesion tests.
- The tungsten bulk resistivity for a 68 nm thick film was 20 $\mu\Omega$cm indicating the formation of α-W.
- SEM micrographs revealed that the oxide/tungsten interface was very smooth.

A good correlation was found between the linewidth of the Mo mask and the tungsten line width. A disadvantage of this technique is that a high dose is needed before tungsten growth starts to occur. With an implant current of 100 μA and a dose of 5E16 at/cm^2, a 6" wafer needs about 250 minutes of implantation time!

8.2.2 Tungsten Growth Using Tungsten Implants

Thomas et al.[205] studied the effect of tungsten implants in oxide on tungsten nucleation. They found that a minimum dosage of $1x10^{16}$ at/cm^2 is required to initiate tungsten nucleation. This is about a factor of 5 lower than the minimum dosage needed in the case of silicon implants which is a clear advantage in terms of implantation time. An additional advantage of the tungsten implants was that the adhesion of the tungsten films was excellent in comparison with Si implants.

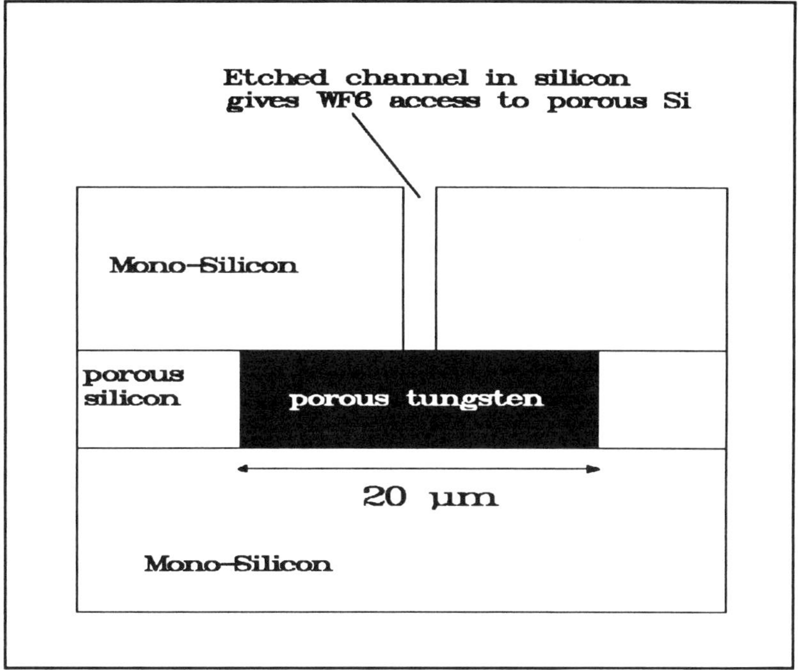

Figure 8.3 By exposing porous silicon via a channel through mono-silicon to an WF$_6$ ambient, buried tungsten islands can be formed underneath mono-silicon.

8.3 BURIED TUNGSTEN

Buried conductive layers underneath mono-crystalline silicon can have interesting opportunities for radiation hardness and multilevel interconnect systems. The basis for a buried tungsten layer is the fact that under appropriate anodic etching conditions, mono-crystalline silicon can be converted into porous silicon. By proper processing, porous silicon can be formed in mono-crystalline silicon substrates. Because the atomic density of Si is about twice that of tungsten, it is expected that exposure of porous silicon to WF$_6$ will result in the formation of porous tungsten. Thus this conversion can potentially proceed over long distances. Tsao et al.[206] investigated the conversion of porous silicon into porous tungsten and found that the conversion rate was diffusion controlled. One would expect this for such a porous structure. The diffusional speed of WF_6 into the pore is rate determining, a situation analogous to the contact fill as discussed in chapter II. The as-deposited tungsten was found to consist of β-W. However, a heat treatment could convert this into α-W giving a resistivity of the porous film in the range of 400 to 4000 $\mu\Omega$cm. This resistivity is considerably higher than that of CVD-W but understandable with regard to the porous structure. The authors were able to grow 20 μm in a lateral direction (see figure 8.3). The structure shown in figure 8.3 is just an example. Other structures involving oxide isolation are also possible.

8.4 ALTERNATIVE DEPOSITION TECHNIQUES

Basically two alternative activation techniques have been reported to deposit tungsten other than by thermal activation. These are activation by aid of a gas discharge (plasma enhanced or PECVD) or by optical activation (photo -mostly laser- enhanced or LCVD). The advantage of these techniques is that the substrate temperature can be relatively low which might be of importance for future developments. In the next two sections we will discuss both PECVD and LCVD.

8.4.1 Plasma Enhanced CVD-W

One approach which can be taken for PECVD is the parallel plate reactor (see figure 8.4). The plasma can be created by either an RF or a DC discharge. When pure WF_6 is used only etching of tungsten will occur according to the gas phase reaction:

$$e + WF_6 \text{ -------> } WF_{6-x} + xF + e \qquad (8.3)$$

which gives atomic fluorine followed by:

$$xF + W \text{ --------> } WF_x \qquad (8.4)$$

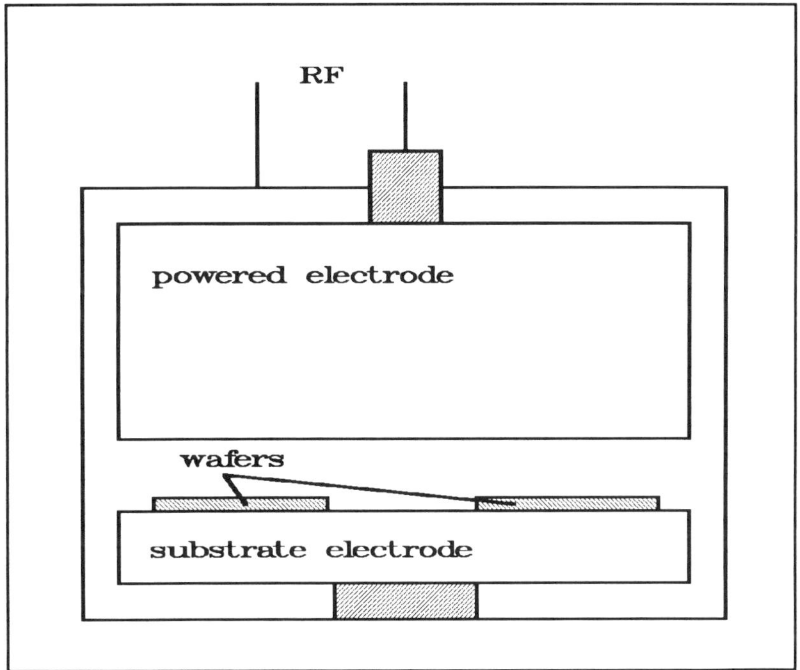

Figure 8.4. Schematic representation of a parallel plate reactor.

The addition of hydrogen, however, suppresses the etching reaction and tungsten films are deposited at about 4 nm/min (substrate temperature 350°C, pressure = 200 mTorr, power density = 0.06 Watts/cm2 and a H_2/WF_6 ratio of 3 [Chu et al.[188]]). The bulk resistivity of such tungsten layers can be quite high 30-80 $\mu\Omega$cm. This high bulk resistivity was later found to be due to the formation of β-W (see also chapter III) [Tang et al.[189]]. Anneal, for instance at 1100°C, of the films lowers the resistivity to 7 $\mu\Omega$cm. The stress of the as deposited film is very acceptable at $6x10^9$ dynes/cm^2.

A different approach in PECVD is the excitation of (one of) the reactants by a remote microwave plasma (see the reactor in figure 8.5). It was argued by Tsuzuku et al.[190] that laser activation (see next section) gives generation of atomic hydrogen in the gas phase. This will take away the need for having a catalytic surface to decompose the molecular hydrogen. Microwave plasmas produce atomic hydrogen according to:

Miscellaneous.

$$H_2 + e \text{ --------> } 2H + e \qquad\qquad (8.5)$$

The very reactive atomic hydrogen will diffuse to the surface and react there with WF_6. The activation energy of the reaction was found to be 39 kJ/mole. Low resistivity (8-12 $\mu\Omega$cm) tungsten films were obtained. The advantage of a microwave plasma is that there is no ion bombardment of the substrate or fluorine generation in the gas phase. These factors are thought to be responsible for the appearance of β-W in parallel plate type of plasmas (see above). Another disadvantage of reactions enhanced by an ion bombardment is that the step coverage will degrade in small features since the ion flux will be shielded and therefore be less dense at the side walls (see figure 8.6).

8.4.2 Photo Enhanced CVD-W

In this section we will describe two forms of LCVD using laser beams. The deposition rate enhancement can be by:

i) A homogeneous photo activation of the reactant(s) in the gas phase or, alternatively, by

ii) A local heating of the surface by the laser beam. In this case direct writing of tungsten lines is possible.

Deutsch and Rathman[193] showed that using an ArF laser beam parallel to the substrate, tungsten depositions can be obtained in a H_2/WF_6 gas mixture at 8 Torr. The substrate temperature was varied between 200 and 440°C. At 440°C, the bulk resistivity was about 17 $\mu\Omega$cm. However, below 350°C, high resistivity (100-300 $\mu\Omega$cm) β-W is obtained. By a heat treatment at 650°C in H_2 the β-W phase can be converted into the α-W phase. The activation energy found for the photo reaction was 40.7 kJ/mole.

Tsuzuku et al.[195] also used an ArF laser with the beam parallel to the substrate (10 mm above). While varying the substrate temperature between 350 and 450°C, tungsten depositions were obtained in a H_2/WF_6 mixture. In their case no β-W was reported. The reaction orders obtained were one for hydrogen and 1/2 for WF_6 according to the rate expression:

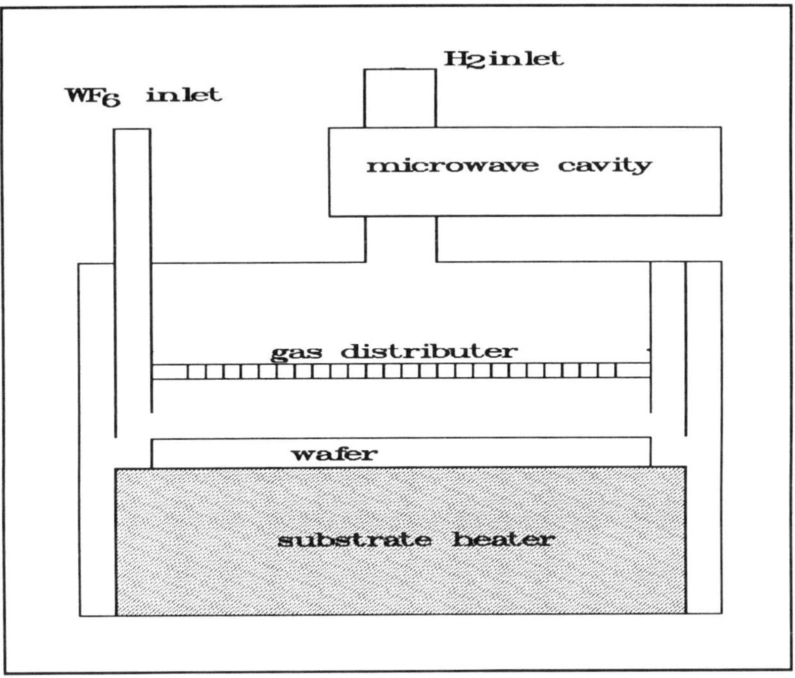

Figure 8.5 Schematic representation of a reactor in which atomic hydrogen can be generated by a micro-wave plasma.

$$\text{Dep. rate} = k\,[P_{WF6}]^{1/2}\,[P_{H2}]^{1} \qquad (8.6)$$

The activation energy for the photo enhanced reaction is 35 kJ/mole which is similar to the value obtained by Deutsch and Rathman. The authors note that this is close to the activation energy of atomic hydrogen diffusion on a tungsten surface (40 kJ/mole). Tsuzuku et al. come to the following proposal for the reaction route:

$$WF_6 + h\nu \text{ -------> } WF_4 + 2F \qquad (8.7)$$
$$F + H_2 \text{ -------> } HF + H_g \qquad (8.8)$$
$$H_g \text{ -------> } H_a \qquad (8.9)$$
$$WF_6 + 6H_a \text{ ------> } W + 6HF \qquad (8.10)$$
$$WF_4 + 4H_a \text{ ------> } W + 4HF \qquad (8.11)$$

Liu et al.[194] used an Ar laser focused perpendicular to the substrate

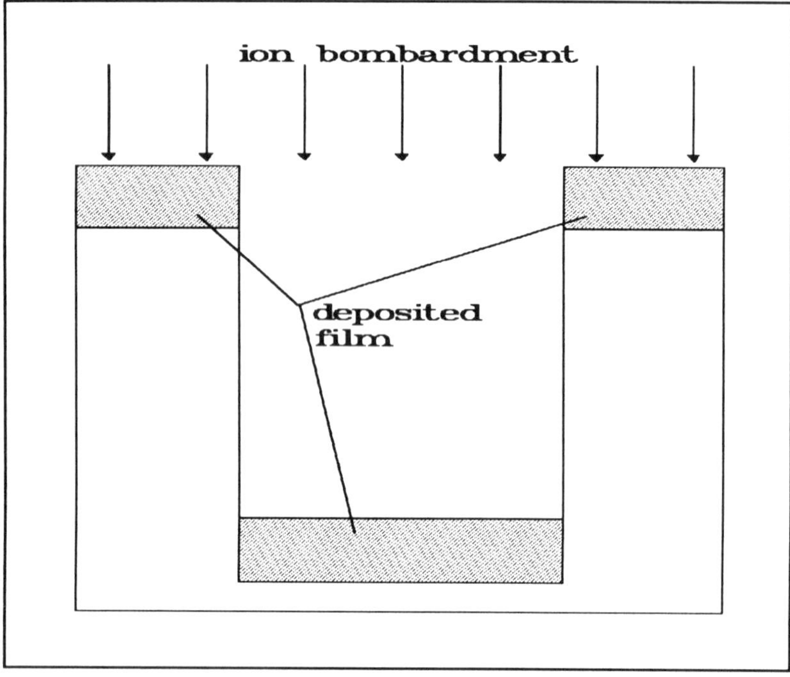

Figure 8.6 The step coverage of a deposition which is enhanced by ion bombardment can be very poor.

and use the local heating effect of the laser to deposit tungsten on Si from only WF_6, at room temperature. Two interesting phenomena important for this direct writing technique need to be mentioned:

a) The chemical reaction rate depends exponentially on the temperature. This results in a considerably narrower linewidth than the actual beam spot size. Depending upon the scan speed and the laser power, linewidths as small as 3 um with a 20 um beam spot size could be obtained.

b) If the dimension of the reacting area becomes small enough, gas phase diffusional transport of the reactants changes from one dimensional to three dimensional (see figure 8.7). The mass flux due to three dimensional diffusion can be much larger than that by semi-infinite one dimensional diffusion. Therefore, deposition rates

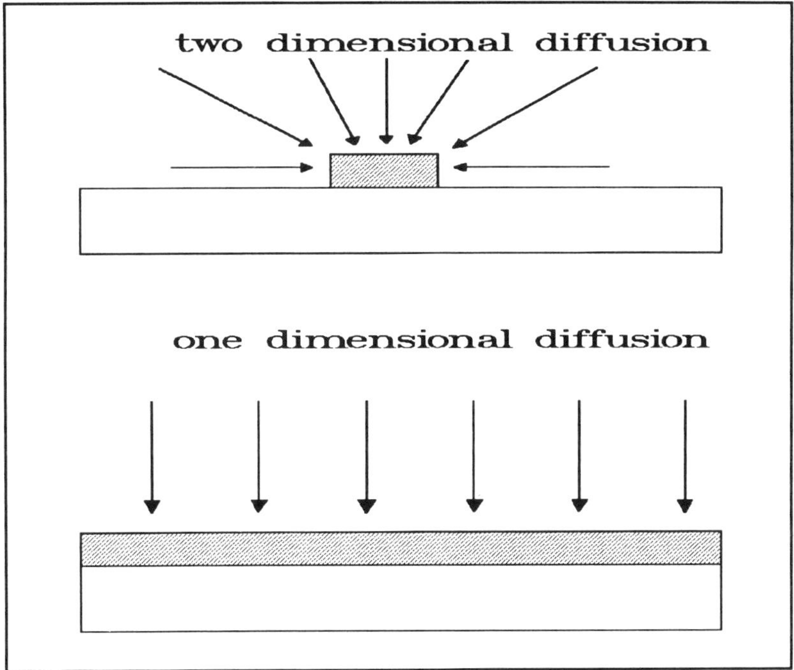

Figure 8.7 2-D diffusion can give much larger mass flux to the surface than does 1-D diffusion.

can be much larger than those obtained for blanket tungsten deposition in the mass transport controlled regime. In this study rates up to 7000 nm/s were obtained!

Only depositions of tungsten on Si substrates were obtained. No tungsten was found on thermal oxide. A complication found was that the tungsten lines were partially silicided in the center.

8.5 ALTERNATIVE PLUG PROCESSES

Several alternative plug processes have been proposed in the literature. Each alternative has its limitations and advantages. The alternatives discussed here are the conversion of poly-Si plugs into tungsten plugs, the SOG/TiW plug and the pillar approach.

8.5.1 Conversion of Poly-Silicon into W

The substitution of (undoped) poly crystalline silicon by tungsten using WF_6 has been studied by Kobayashi et al.[191]. It was found that after a proper pretreatment (=oxidation in H_2O_2/NH_4OH) of the poly-Si very thick (up to 1.5 μm) tungsten could be obtained by reaction with WF_6 at 300°C. The properties of the tungsten films obtained were:

> Resistivity: 70-100 $\mu\Omega$cm
> Auger: oxygen 1-4 at%; Si < 1 at%
> X-ray: α-W
> Density: 13-14 g/cm^3
> Stress: $2x10^9$ dyne/cm^2

This material is without any doubt very acceptable for plug applications. The process flow for a contact to Si fill is then as follows (see also figure 8.8):

1) After contact etching 200 nm TiN is reactively sputtered onto the surface.
2) Poly-Si is deposited at 625°C, such that the contacts are completely filled.
3) Back etching of the poly-Si layer is done using RIE in SF_6.
4) Reforming "chemical oxide" in H_2O_2/NH_4OH.
5) Substitution of poly-Si by tungsten by a WF_6 exposure at 300°C for 30 min. This step is possible because the volume of the replaced Si is about twice that of the formed W. In fact porous tungsten is formed, as can be inferred from the low density mentioned above.
6) Aluminum deposition and alloy step at 450°C in H_2.

An identical procedure can be applied for vias to WSi_x or W. In these cases the TiN layer is not needed since the conversion will stop on the metal automatically. Contact resistivities were measured and were very acceptable: $1x10^{-8}$ Ωcm^2 for the vias and $2x10^{-7}$ for n$^+$ and $4x10^{-7}$ Ωcm^2 for p$^+$ mono-

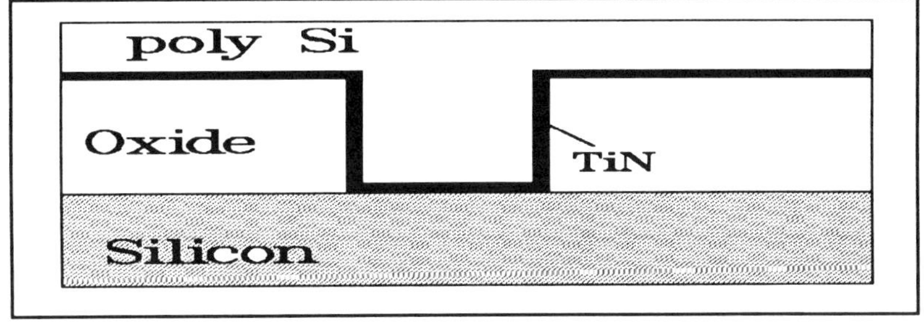

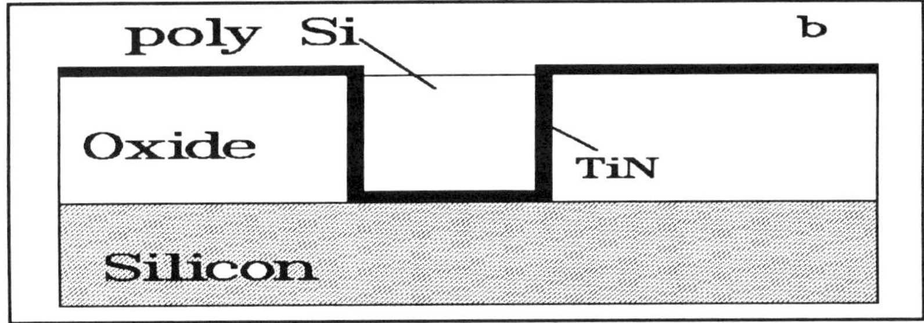

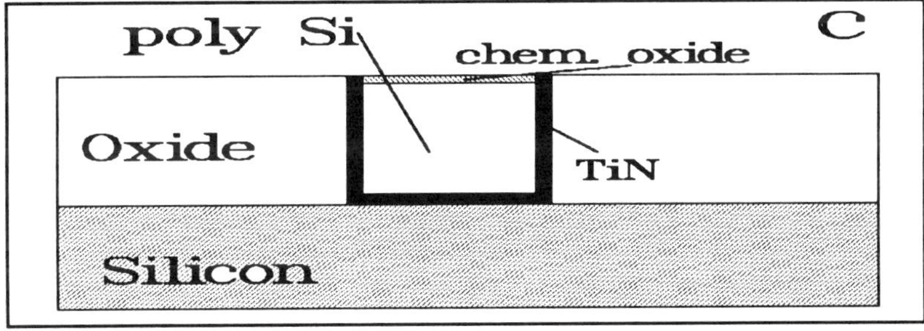

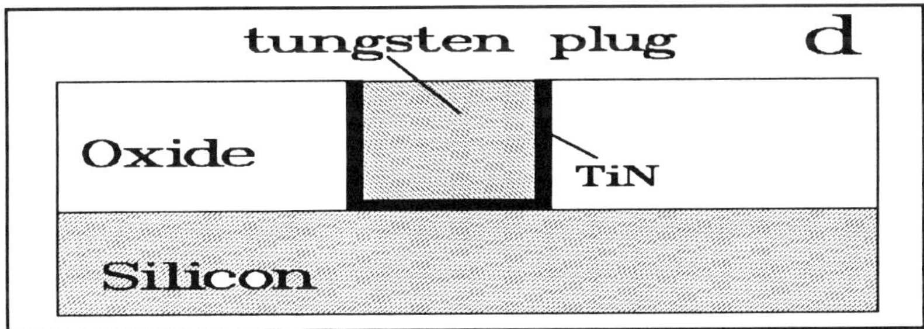

Figure 8.8. Conversion of poly-Si plugs into W plugs (see text).

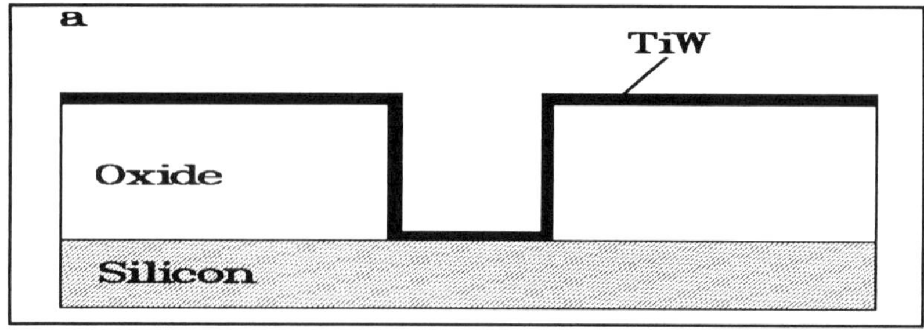

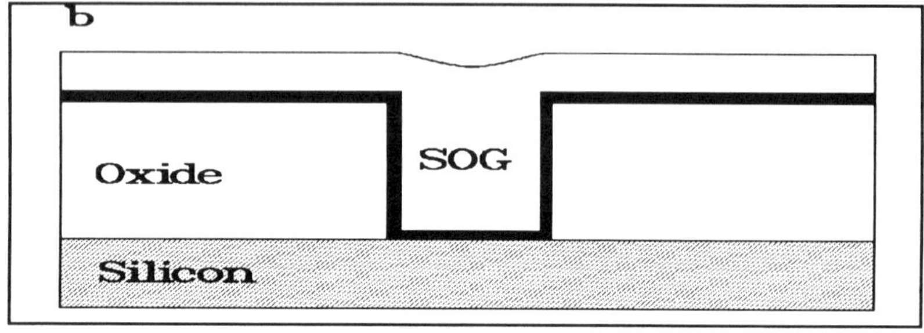

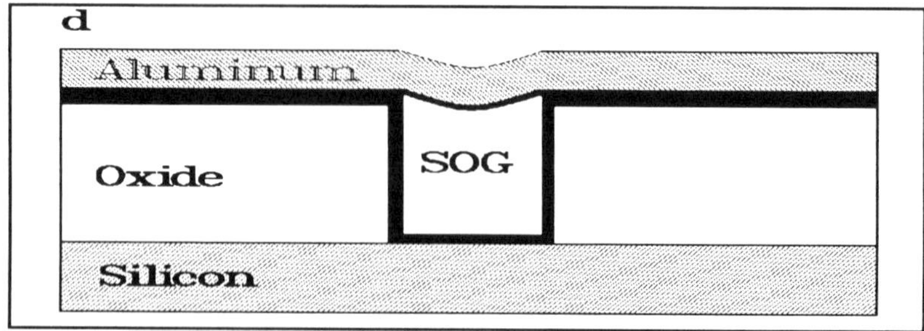

Figure 8.9. The SOG/TiW plug, see text for details.

crystalline silicon.

A similar approach, but now for interconnect applications, was followed by Black et al.[192] to convert laser written poly-Si lines partially into tungsten. The poly-Si lines were exposed for 6 minutes to a WF_6/Ar mixture at 0.625 Torr. At 475°C, 0.4μm thick doped poly-Si lines exhibited a conductivity improvement of up to a factor of 20. RBS and SEM analysis showed that about 100 nm of tungsten was deposited on the poly-Si and that the resistivity of the tungsten film was about 10 $\mu\Omega$cm. The results here, as in the poly-Si plug conversion case , were rather dependent how exactly the poly-Si was pretreated in terms of residual oxide thickness.

8.5.2 SOG/TiW Plugs

Another approach for contact fill is the SOG/TiW plug [Schmitz et al.[187]]. The process flow is represented in figure 8.9. In the first step we see the situation after contact etching and TiW sputter deposition. The step coverage of the TiW deposition is about 50% in a contact of one micron diameter and a depth of one micron. Therefore, electrical continuity is maintained. In principle a TiN layer could also be used provided the step coverage is in the same range. In the second step Spin On Glass (SOG) is deposited. This is the most critical step since the SOG tends to crack in the contact if not cured (baked) in the proper way. In this study the SOG was deposited in two separate steps with intermediate cures. Then an etch back of the SOG was performed using a CF_4/CHF_3 plasma such that it stopped at the TiW layer. Finally, in step IV, TiW and Al are sputtered and patterned.

The electrical continuity is provided by the TiW layer at the side wall of the contact. When we assume a bulk resistivity of 150 $\mu\Omega$cm for the TiW, the total resistance of the TiW in a contact with a diameter of 1 μm and 1 μm deep will be about 6 Ω (assuming 0.1 μm thick TiW at the side wall). This value is probably acceptable for contacts to silicon where the contact resistance will be in the range of 20-50 Ω per contact. For vias, however, where the contact resistance is in the range of 0.1-0.5 Ω this is unacceptable.

Miscellaneous.

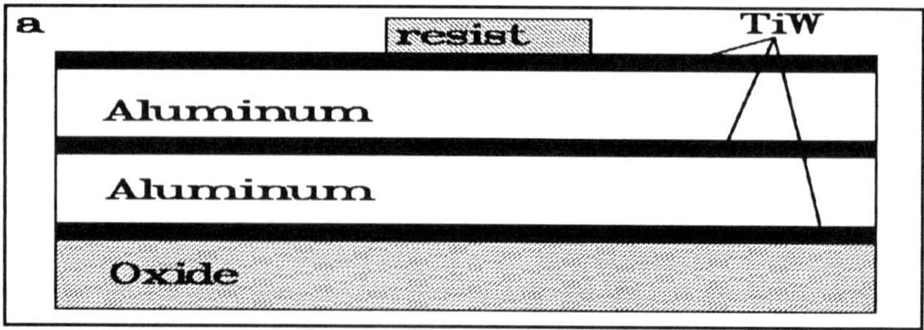

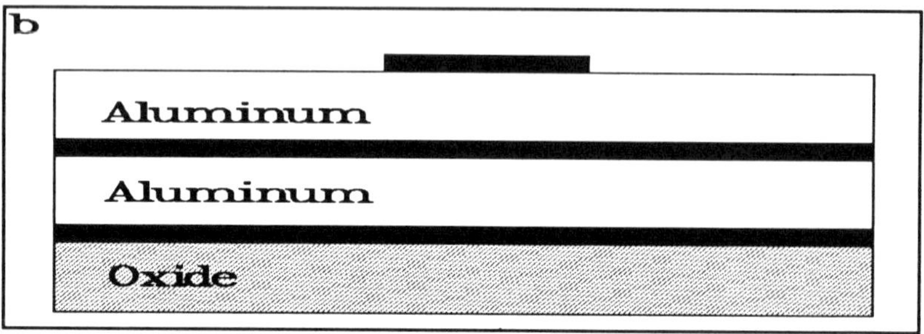

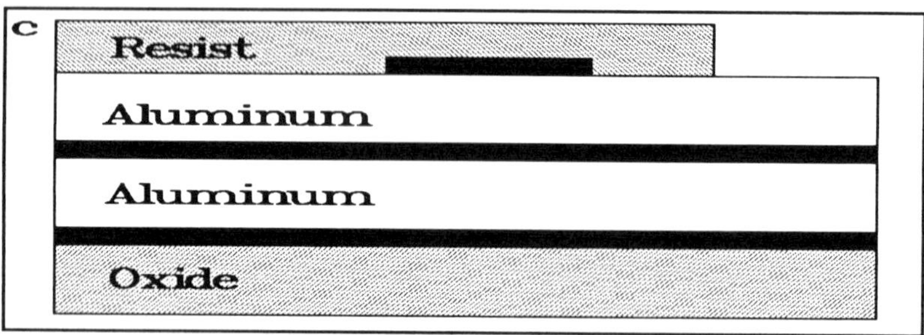

Miscellaneous.

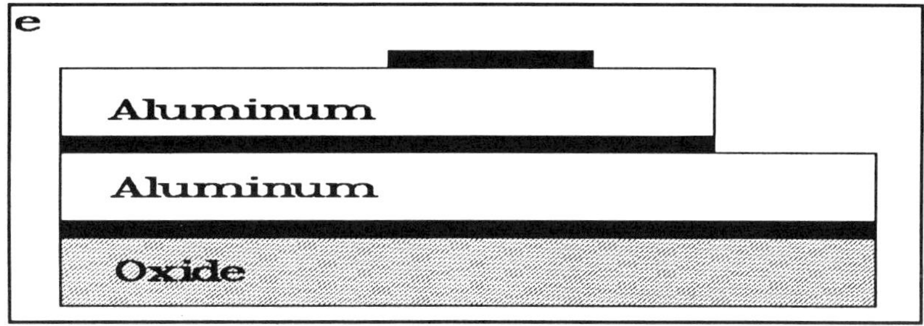

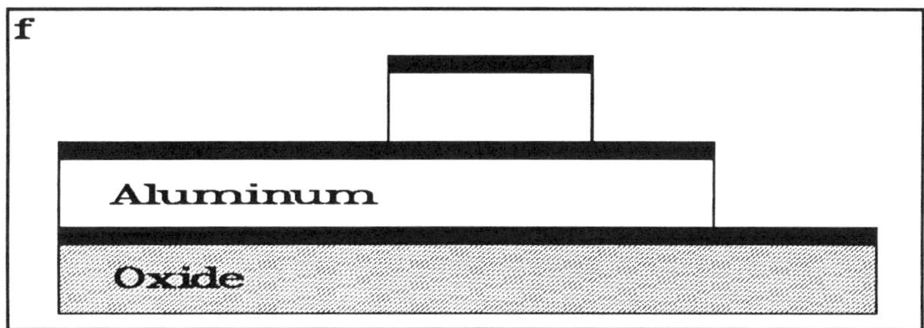

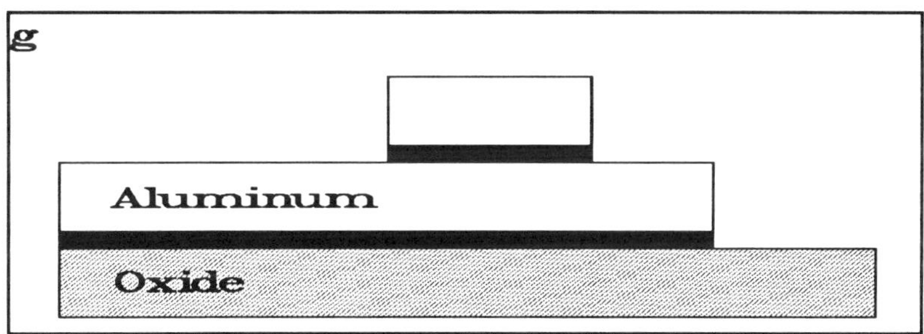

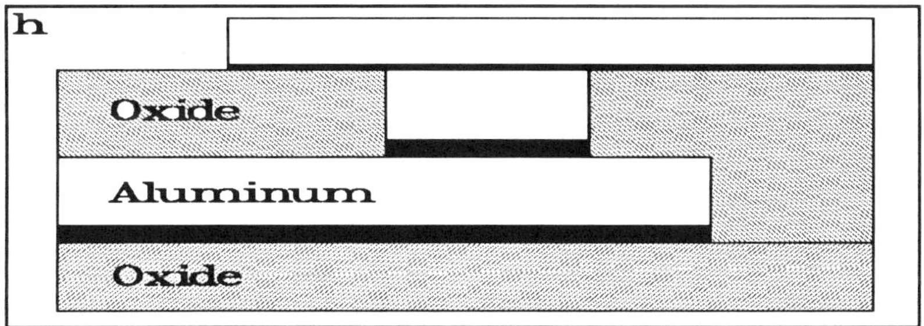

Figure 8.10. Pillar technique, see text for details.

8.5.3 The Pillar Approach

As will be clear from the following explanation, the pillar process has the same advantages, in terms of Si real estate, as has the tungsten plug process (see chapter I). The process steps have been described by Welch et al.[184] (see also Yeh et al.[185] for pillar variations). In figure 8.10 we see a sketch of the complicated process flow. The starting metal layer consists of two AlCu films sandwiched between three TiW layers. On the stack a photoresist is spin coated in which the inverse via mask is printed (figure 8.10.a). The TiW layer is etched in a fluorine based plasma with good selectivity towards the AlCu layer after which the resist is stripped (figure 8.10.b). Now another photoresist layer is applied and the first interconnect mask is exposed (figure 8.10.c). In a chlorine etch the top AlCu layer is etched. The etch will stop on the second TiW film (figure 8.10.d). The pattern in the upper aluminum film is now used to pattern (using a fluorine etch which stops on the lower aluminum film) the middle TiW layer and the resist is stripped (figure 8.10.e). Both the pillars and the first interconnect aluminum are now etched using the TiW as a hard mask in a chlorine etch which will stop on the lower TiW layer (figure 8.10.f). Finally the lower TiW film is patterned using the lower aluminum as a mask. The TiW on top of the pillar will also be removed in this last step (figure 8.10.g). After this a planarization step has to be carried out using the Resist Etch Back technique (figure 8.10.h). Critical in this step is that all the pillars are exposed. Once this is accomplished the second interconnect film can be deposited and if necessary the same pillar technique can be applied.

A disadvantage of this approach is that it obviously will not work for the fill of contacts to silicon (however, see Yeh et al.[185]). This implies that two plug techniques are necessary to overcome the reliability issues, one (blanket) for the contacts and one (pillars) for the vias.

170

CHAPTER IX

CHEMICAL VAPOR DEPOSITION
OF TUNGSTEN SILICIDE

9.1 INTRODUCTION

The previous part of this book covers the chemical vapor deposition of tungsten. However, there are several reasons to include a discussion about the chemical vapor deposition of tungsten silicide (CVD-WSi$_x$). These are:

- The chemistries in use for CVD-W can also be used to deposit WSi$_x$, al be it that the deposition conditions slightly change.

- Nearly identical equipment as used to perform CVD-W can be used (and as a matter of fact has already been in use for years) to deposit WSi$_x$.

- Both materials can be part of a high speed multi-level interconnection system. WSi$_x$ has already been in use for years world-wide in different types of IC's in large volume production. Tungsten is now close to this phase.

- Another reason is that CVD-WSi$_x$ based on the SiH$_4$/WF$_6$ chemistry approaches certain limitations as the feature size of IC's becomes smaller than one micron and as gate oxides become thinner and thinner (say below 200Å). It will be shown in later sections, that a deposition chemistry based on SiH_2Cl_2/WF_6 can overcome these limitations.

In 1983 an extensive review of silicides for IC's applications was published by Murarka[86]. This work focused mainly on sputter techniques as the deposition method. As we will see, after 1983 the CVD technique became the most popular method for polycide applications. Much literature on the technique itself and on the film properties has been published. It seems appropriate here to summarize from the literature the most significant results reported after 1983 on CVD-WSi$_x$.

In this chapter we will briefly mention the use of WSi$_x$ for polycide structures to overcome line delay problems and the attractiveness of the CVD technique to deposit WSi$_x$. In addition, we will elaborate on the two pertinent chemistries for CVD-WSi$_x$ namely SiH_4/WF_6 and SiH_2Cl_2/WF_6.

One remark about using WSi$_x$ rather than WSi$_2$ is in order. We will see that seldom is pure WSi$_2$ deposited but rather a mixture of WSi$_2$ and Si. The composite film is then more accurately described by WSi$_x$. From a chemical point of view this notation is meaningless and does not imply a real compound with a silicon over tungsten ratio of x!

9.2 WSi$_x$ FOR POLYCIDE APPLICATIONS

In VLSI-MOS based circuits, where doped poly-crystalline silicon (poly-Si) is used as the gate electrode, the performance of the IC is limited by the RC time delays in the poly-Si runners (see for detailed discussions Murarka[86], Sachdev et al.[212]). The main reason for this delay is the high resistivity (500 $\mu\Omega$cm) of n+ doped poly-Si. In addition, the trend to a larger die size worsens the problem because these poly-Si lines become longer.

Several approaches to decreasing the RC value of this gate electrode can be followed, such as replacing the entire poly-Si gate by a metal or a silicide. This approach, however, is not easily retrofitable into existing processes. One reason is that the well characterized, high quality gate oxide/poly-Si interface is no longer retained when the poly-Si electrode is replaced by another material. Therefore, the polycide (=poly-Si/silicide stack) solution has gained extensive popularity and has been studied in great detail. In this approach, the poly-Si is cladded with a low resistivity silicide such that the sheet resistance of the stack becomes about 3 $\Omega/\square$ (note without silicide this will be about 20-30 $\Omega/\square$). In figure 9.1 the simplified process flow of a polycide process is sketched.

What requirements have to be fulfilled in order to have a truly retrofitable and acceptable solution?

- The sheet resistance of the polycide stack should be as low as possible.

- The silicide must provide low ohmic contacts to other levels of metallization.

- The patterning of the polycide stack should give no additional problems compared with the patterning of poly-Si.

- The silicide should give an acceptable oxide quality upon oxidation and should be capable of withstanding high post process temperature steps.

- The silicide should resist attack by chemicals used in IC processing such as HF solutions.

Several silicides are able to fulfill most or all of these conditions. The first silicides studied with respect to the polycide application were MoSi$_2$ and WSi$_2$ (see Crowder[214]). Later TaSi$_2$, TiSi$_2$ and CoSi$_2$ were also studied, the latter two especially for use in the salicide processes. WSi$_2$ became the first choice for use in polycide processes. This was not only because it met all the requirements listed above but also because a suitable deposition technique was developed (*vide infra*).

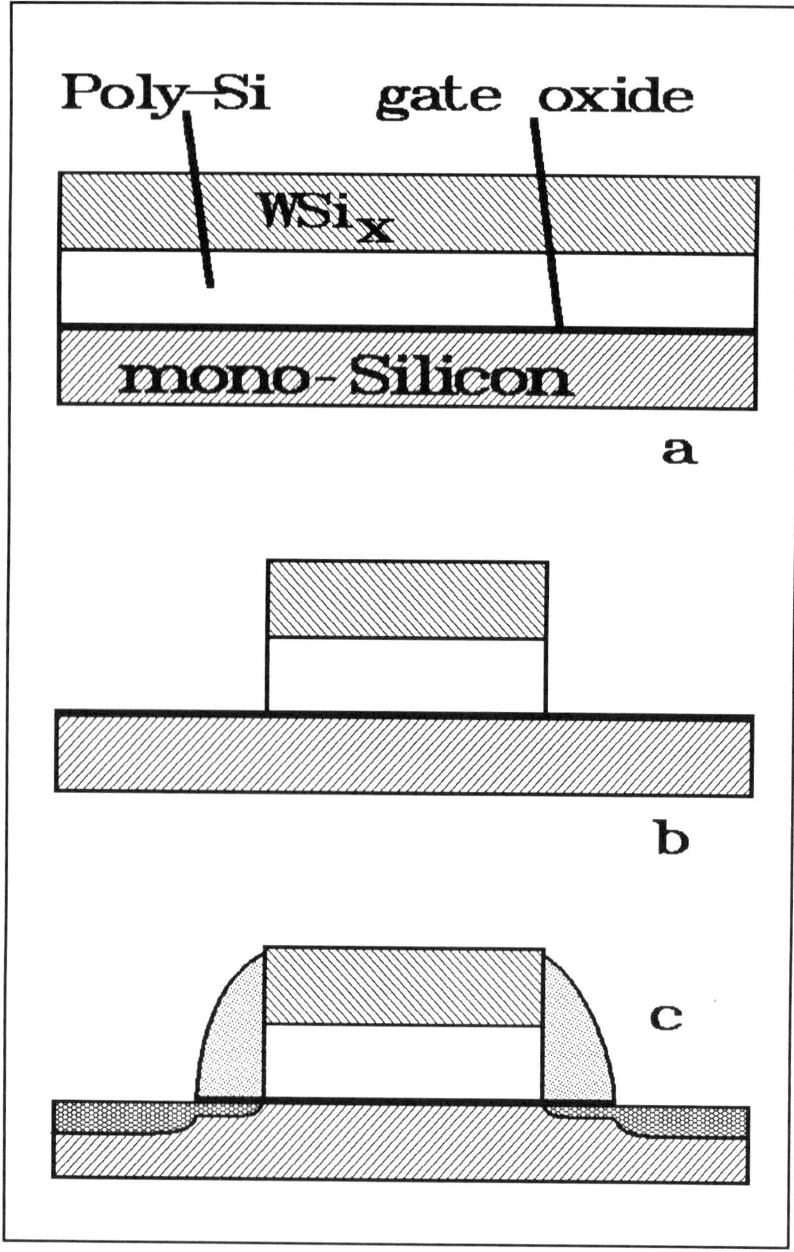

Figure 9.1. MOS gate structure: a) after deposition of poly-Si and the silicide; b) after patterning the polycide stack; c) after S/D implantations and spacer formation.

174

Another possible application of tungsten silicide is that of local interconnect [Mihara et al.[215]]. We will refrain from a further discussion of this application.

9.3 SILICIDE DEPOSITION METHODS

For some time, difficulties with the deposition of silicides were preventing successful large volume implementation. Several techniques have been tried:

- Evaporation. The problem here is how to get a repeatable composition of the as deposited film. Because of the large vapor pressure difference between silicon and tungsten (at 3000°C!) a single source cannot be used. The lack of control of the Si/W ratio in the film using two sources makes run to run repeatability difficult to maintain. Another disadvantage of the e-beam technique is the extremely poor step coverage ("line of sight" profile, see chapter II). However, for specific applications such as lift off, this could be an advantage.

- Sputtering. To some extent we have the same problem here as with evaporation, namely, difficult control of the composition (and purity) of the as deposited film. High purity and homogeneous composite targets are expensive and therefore not very attractive in high volume production. Using two targets, one for silicon and one for tungsten, can give good quality films, but composition is more difficult to control. Again the step coverage can be rather poor as well. An advantage of the sputter technique is that the poly-Si can be cleaned using a "soft" sputter etch prior to the silicide deposition. Another advantage is that the silicide can be formed by sputter depositing only the metal and then reacting the metal film with the underlying silicon in a subsequent step. This scheme has been shown to be quite promising and is called a salicide (self aligned silicide) process.

- CVD. In 1983 it was shown by Brors et al.[217] that a good quality WSi$_x$ film could be deposited using the SiH$_4$/WF$_6$ chemistry in a cold wall reactor. Among the advantages of this technique are: a) no need for high vacuum equipment to obtain a high purity film; b) acceptable throughput; c) very good step coverage as compared with physical deposition techniques and d) very good wafer to wafer and run to run uniformity. A drawback of this procedure was found in later years, namely, the high fluorine content of the as-deposited films (see section 9.7).

For a more extensive comparison of these techniques see Ahn et al.[213] and Crowder[214].

After 1983, CVD-WSi$_2$ became because of the above mentioned reasons popular and is now worldwide in use in large volume production (almost exclusively in polycide applications). Besides CVD also sputtered WSi$_2$ is still in use in production. In the following sections we will elaborate on the CVD technique.

9.4 CVD OF WSi$_x$

There are several methods possible with the CVD technique to come to WSi$_x$ films as we will mention below. Lehrer and Pierce[218], have described an interesting approach in which they do a sequential deposition of Si and W. The chemistry used was:

$$SiH_4 \ \text{-----} > \ Si \ + \ 2H_2 \qquad (9.1)$$

and

$$WCl_6 \ + \ 3H_2 \ \text{-----} > \ W + 6HCl \qquad (9.2)$$

Both reactions were carried out in a cold wall reactor at 600°C and at atmospheric pressure. First, 400nm of silicon was deposited followed by 65nm of W. After an anneal at 1000°C in Ar for 10 min a thin film resistivity of about 100 $\mu\Omega$cm was obtained. This high value was due to oxygen incorporation during the W film deposition.

176

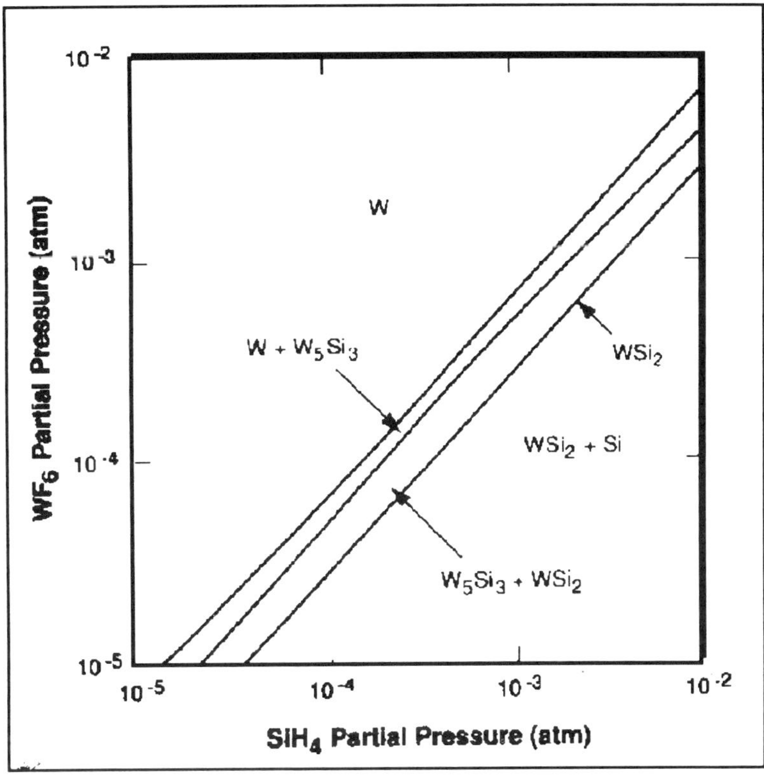

Figure 9.2. Equilibrium CVD phase diagram for the system W-Si-F-H-Ar at 1000K. [From Bernard et al.[221], reprinted with permission of Solid State Technology].

Akitmoto and Watanabe[219], have reported the deposition of $W_x Si_{1-x}$ mixtures by plasma enhanced CVD (PECVD). Depending on the flow ratio of WF_6/SiH_4, x can be varied from 0.04 to 0.99. The deposition was carried out at a substrate temperature of 230°C. It is interesting that both the as-deposited and annealed (1100°C, 60 min., N_2) films, with x values of 0.45 or less, did not exhibit any X-ray diffraction pattern due to a WSi_2 phase. From their data it can be estimated that a near stoichiometric disilicide film would have a resistivity of about 200 $\mu\Omega$cm after anneal. The authors claimed that specular films were obtained.

Dobkin et al.[220] have performed CVD of WSi_x under atmospheric conditions using a gas mixture of $WF_6/Si_2H_6/H_2/N_2$ at 300°C. Films of

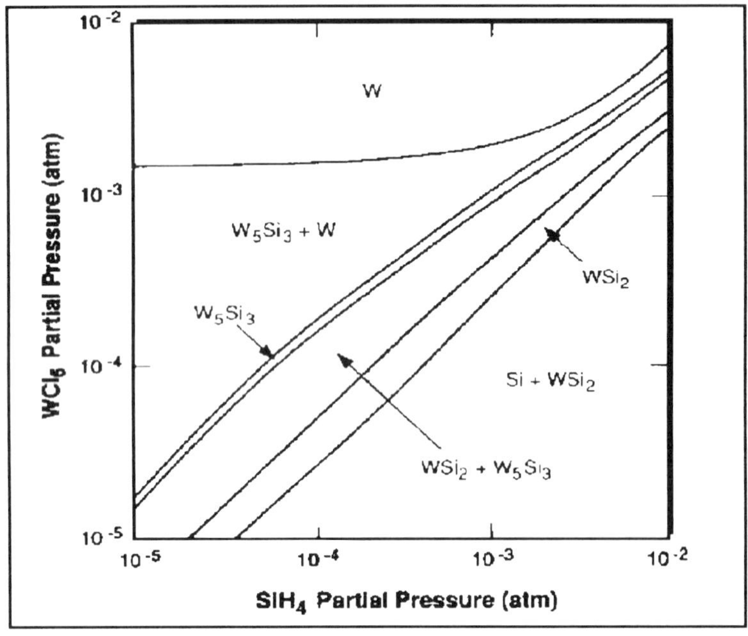

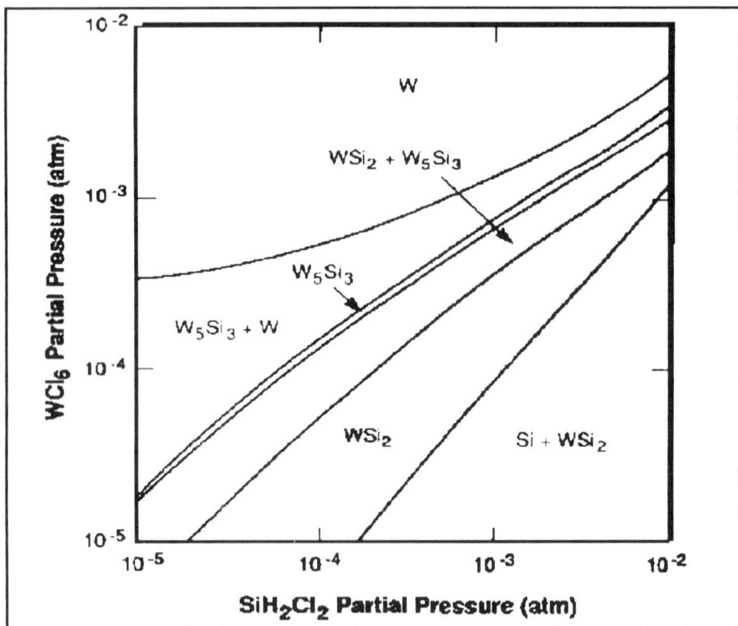

Figure 9.3. CVD phase diagram for WCl$_6$/SiH$_4$ (top) and WCl$_6$/SiH$_2$Cl$_2$ (bottom) at 1000 K. [Bernard et al.[221], reprinted with permission of Solid State Technology].

various compositions were obtained. Unfortunately, no further details such as resistivity after anneal or oxidation were disclosed.

In 1983 Brors et al.[217] proposed to deposit WSi$_x$ in a cold wall reactor using the SiH$_4$/WF$_6$ chemistry. This appeared to be a very successful approach as it is still in use in the industry. Good quality silicide could be obtained which gave post anneal resistivities as low as 35 $\mu\Omega$cm. In section 9.5 we will discuss this chemistry and the material properties in much more detail. Before we do this it is wise to consider first some CVD phase diagrams in order to understand better the results obtained from CVD-WSi$_x$.

Bernard et al.[221], have calculated the CVD phase diagrams for several silicide systems. These phase diagrams are based on a calculation which minimizes the Gibbs energy for a certain fixed amount of chemical species. Although very helpful, we should keep in mind the following limitations:

- First a choice has to be made as to what species will be included in the calculations.

- Of all the included species, thermodynamic data (enthalpy, entropy, heat capacity etc.) has to be found and fed into the calculation. This can be very difficult since not all of this data is always known or correct. The quality of the outcome of the calculation depends of course on the quality of the input data.

- The calculations represent the situation once equilibrium (that is minimum free energy) is reached. The **speed** (i.e. in our case the deposition rate) with which this condition is reached cannot be predicted. Moreover, a cold wall reactor is by definition not in equilibrium since the temperature varies spatially. The reactor can, however, reach a steady state (that is for a given input of reactants the spatial concentrations will not change in time). Note, however, that a steady state is **not** an equilibrium.

Thus, especially in cold wall reactors, deviations from thermodynamic calculations can be found (see also chapter III). Nevertheless, thermodynamic considerations can give valuable insight into the possibilities

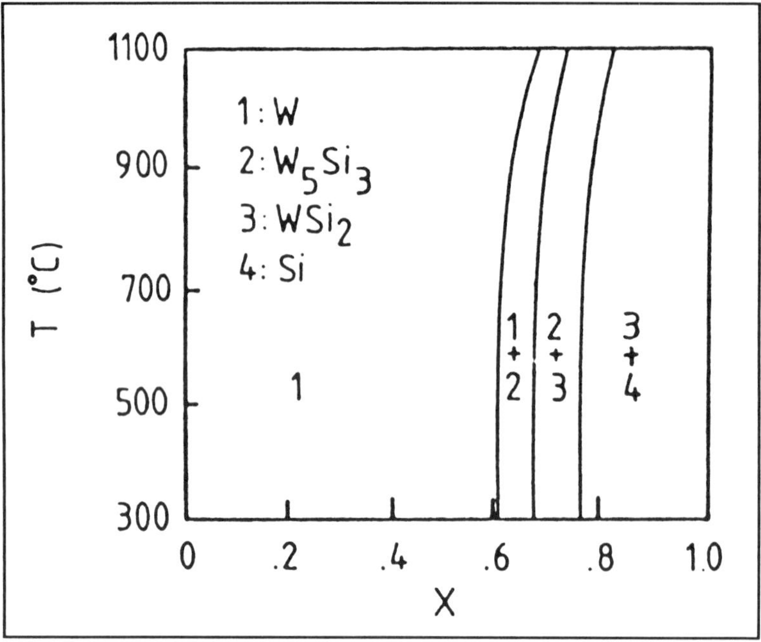

Figure 9.4. Equilibrium CVD phase diagram for SiH$_4$/WF$_6$ at 0.5 Torr. X = silane mole fraction. [From Zhang et al.[222], with permission].

of a certain chemistry. Consider for example figure 9.2. We see that the existence area of stoichiometric WSi$_2$ is limited to a very narrow process range. Fortunately, as pointed out by several studies [Rode et al.[223], Murarka[86], Brors et al.[217]] in practice a silicon rich mixture is preferred (see section 9.8). We see that according to the phase diagram this is possible with the SiH$_4$/WF$_6$ chemistry. Also, the composition of the Si-WSi$_2$ mixture is dependent on the **ratio** of the reactants. By using a different chemistry for instance, replacing WF$_6$ by WCl$_6$ or SiH$_4$ by SiH$_2$Cl$_2$, the phase diagram can change substantially (see figure 9.3). For both the WCl$_6$/SiH$_4$ and the WCl$_6$/SiCl$_2$/H$_2$ chemistries, the existence range of WSi$_2$ is increased. Unfortunately, no good thermodynamic data for the WF$_6$/SiH$_2$Cl$_2$ system is available [Bernard et al.[221]].

A good example of the limitation of this approach can be found in figure 9.4. According to this phase diagram the composition of the as-

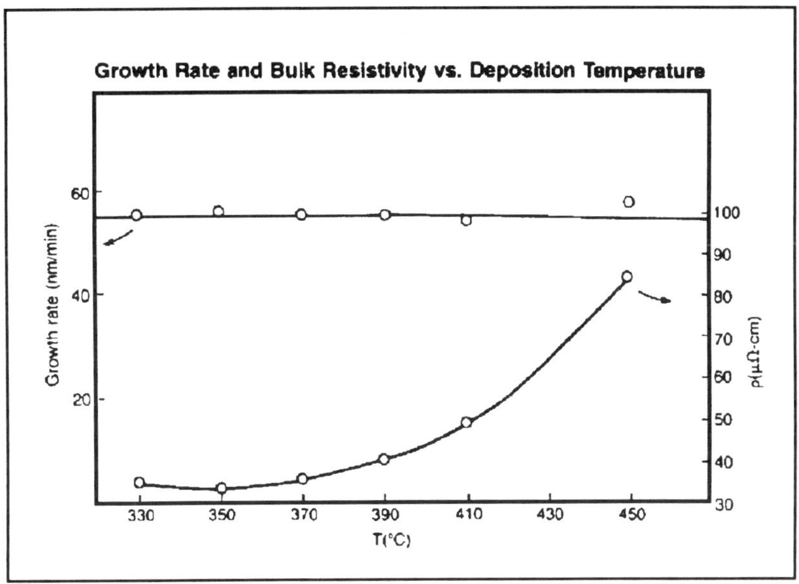

Figure 9.5. The deposition rate of WSi$_x$ and thin film resistivity (after anneal) as a function of the temperature. [Brors et al.[224], with permission, © 1984 Semiconductor International].

deposited film in the silicide regime is not very temperature dependent. In reality the silicon content is very temperature sensitive (vide infra) above about 350°C. In fact, we have here the same problem as already encountered in the discussion of the selective W chemistry in chapter III. Thermodynamically, silane is an unstable compound and will therefore decompose into silicon and hydrogen (in the model). However, reaction 9.1 will not occur at temperatures below say 500°C although part of the silane may be catalytically decomposed by the action of the tungsten atoms in the growing film. Thus in practice, reaction 9.1 will play a larger role at higher temperatures and will cause an increase in the silicon content of the film with increasing temperature.

From thermodynamic calculation of the composition of the gas phase, Rode et al.[223] came to the following reaction stoichiometry for the SiH$_4$/WF$_6$ chemistry:

$$2WF_6 + 7SiH_4 \text{ -------> } 2WSi_2 + 2SiF_4 + 14H_2 \qquad (9.3)$$

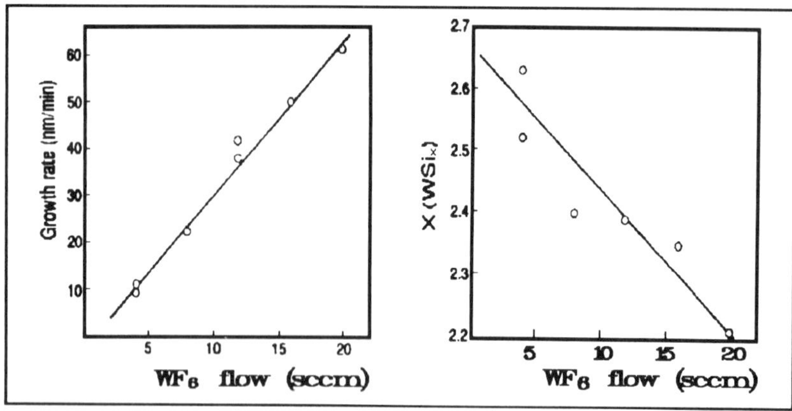

Figure 9.6. Influence of WF$_6$ flow rate on deposition rate and composition. [Brors et al.[224], reprinted with permission, © 1984 Semiconductor International].

This reaction proceeds together with reaction 9.1. Reaction 9.1 and 9.3 can account for 99% of the silane consumption. The overall reaction for the dichlorosilane chemistry (DCS) is:

$$2WF_6 \ + \ 10SiH_2Cl_2 \ \text{-------> } 2WSi_2 \ + \ 3SiF_4 \ + \ 3SiCl_4 \ + \ 8HCl \ + \ 6H_2 \quad (9.4)$$

9.5 CVD-WSi$_x$ BASED ON SiH$_4$/WF$_6$ CHEMISTRY

9.5.1 Deposition Process and Film Properties

In this section we will discuss the process in terms of reaction kinetics, film composition and film properties. Important to note is that the degree of utilization of WF$_6$ under typical deposition conditions (ie. 50 nm/min and 2 sccm WF$_6$ per 6" wafer) is at least 50%. Thus the reactor is very likely to run in a depletion or feed controlled mode. As a result we expect the deposition rate to depend strongly on the WF$_6$ total flow and be relatively insensitive to temperature variation.

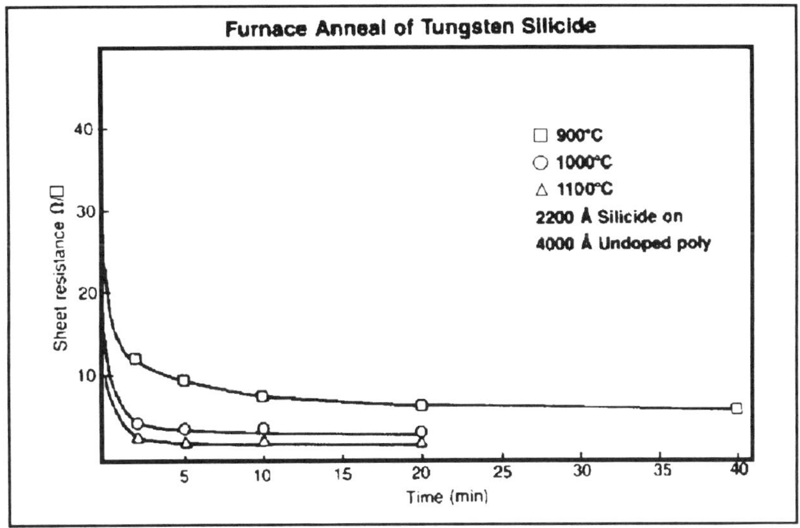

Figure 9.7. Sheet resistance versus anneal temperature and time. The as-deposited Si/W range: 2.6-2.8. [Brors et al.[224], reprinted with permission, © 1984 Semiconductor International].

Table 9.1

Grain size and dominant crystal phase at different anneal temp.

Temperature °C	grain size nm	dominant structure
as depos.	3	no reflect.
500	10	hexagonal
600	20-30	tetragonal
800	40-50	tetragonal
1000	100	tetragonal

Data from Saraswat et al.[225].

As to mentioned above, one of the first characterizations in a coldwall system was done by Brors et al.[217,224] in 1983 and in 1984. It can be seen in figure 9.5, that the deposition rate is indeed insensitive to temperature variations in the range 330-450°C. An explanation for the increase in the thin film resistivity (after anneal) with higher deposition temperature will be given in section 9.5.3. In figure 9.6 we see that both the deposition rate and the composition of the film depend very strongly on the WF$_6$ flow rate. The ultimate after anneal resistivity of a given film depends strongly on the anneal temperature as can been seen from figure 9.7. At 1100°C resistivities as low as 35 $\mu\Omega$cm were obtained. The time in which the lowest value of the resistivity is reached, is shorter at higher anneal temperatures. A 30 minutes anneal seems to be a sufficiently long time. During the anneal, the silicon content of the film is reduced (see section 9.5.3. for more details). A TEM and X-ray study [Saraswat et al.[225]] gave the following results for the grain size and the predominant lattice structure (see table 9.1) at different anneal temperatures. The recrystallization starts at about 500°C to initially form the hexagonal phase. At 600°C the predominant phase is tetragonal.

The oxidation rate in dry oxygen was found to be very similar to that of <100> silicon (see figure 9.8). This indicates that during oxidation silicon from the underlying poly-Si diffuses to the surface of the WSi$_x$ where it becomes available for oxidation. It has been found [Saraswat et al.[225]] that for practical purposes the dry oxidation rate of CVD-WSi$_x$ atop of poly-Si can be described by:

$$X^2 = B \times t \tag{9.5}$$

where X is the oxide thickness after oxidation time t and B is the parabolic rate constant.

A detailed process characterization was published by Clark[226] in 1988. Using experimental design the deposition process and resultant film properties were studied. Some results obtained in this study are presented in the contour plots in figures 9.9 and 9.10. The centerpoints used in this study were: temperature = 360°C, SiH$_4$ flow = 1200 sccm, WF$_6$ flow = 12 sccm and pressure = 230 mTorr. It was concluded from these contour plots that a fairly wide process window will give deposition rates of about 60

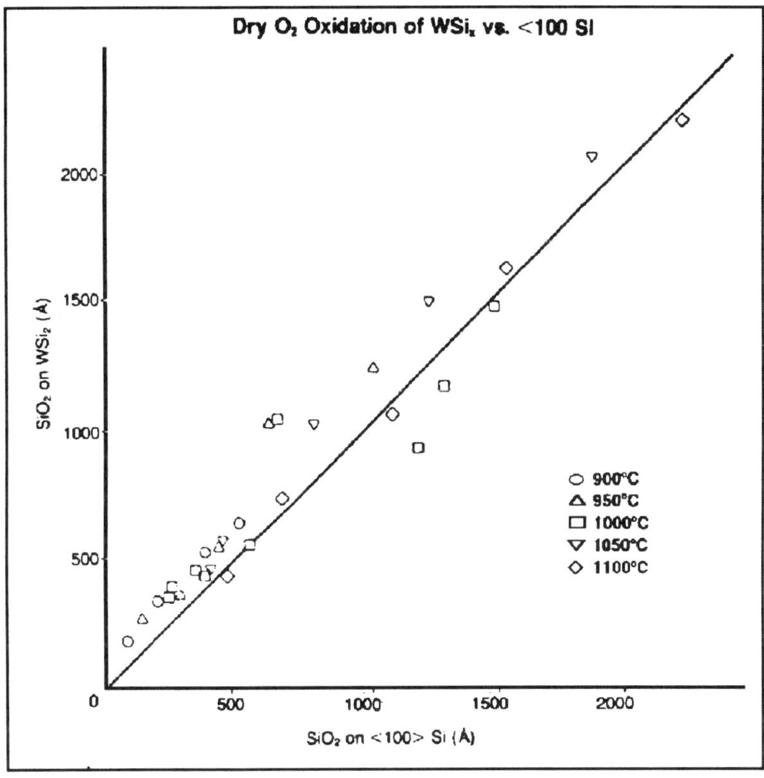

Figure 9.8. The oxidation rate of WSi$_x$ is very similar to that of <100> Si. [Brors et al.[224], reprinted with permission, © 1984 Semiconductor International].

nm/min., and film properties such as annealed resistivity of 75 $\mu\Omega$cm, Si/W ratio of 2.5 and a tensile stress of 1.2×10^{10} dynes/cm^2. This wide window, of course, facilitates the use of this process in IC manufacturing.

The as-deposited thin film resistivity increases with increasing silicon content as is clear from figure 9.11. The post-annealed resistivity follows a similar trend as observed earlier by Brors et al.[224]. The tensile stress of both the as-deposited and annealed film decreases with increasing Si content. It is important to note that the stress after anneal is higher than before anneal. For more details about stress see section 9.8.

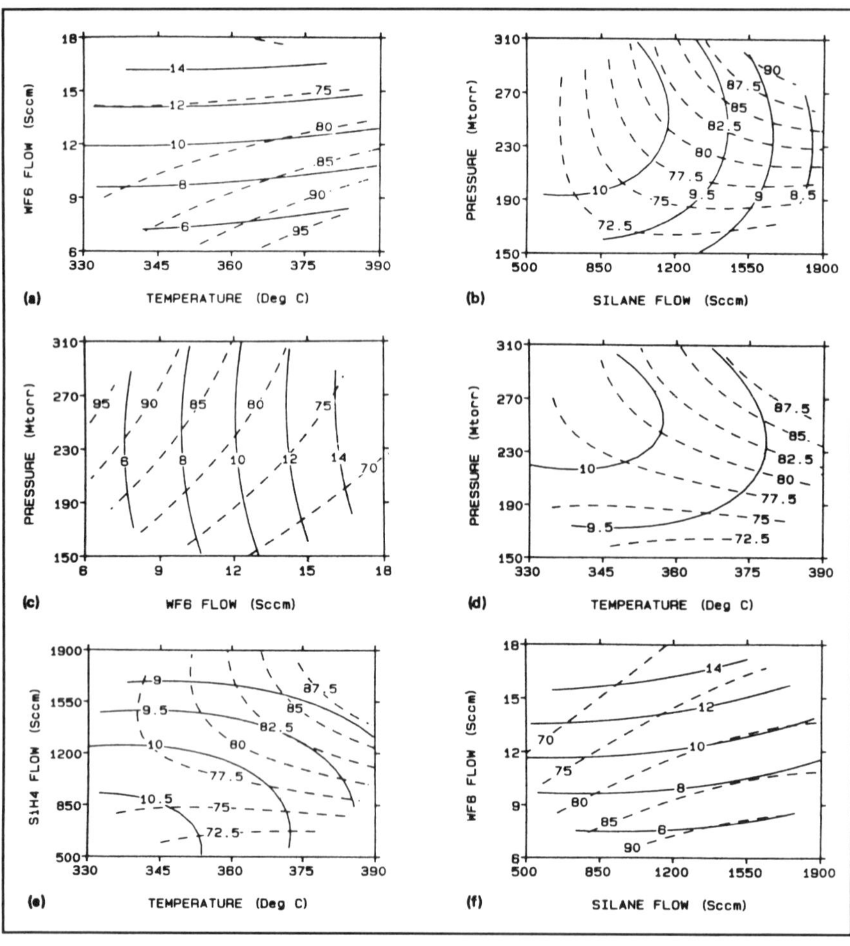

Figure 9.9. Contour plots of dep rate [Å/s (-)] and post-annealed res. [$\mu\Omega$cm (--)]. Other variables at their centerpoints, see text. [Clark[226], reprinted with permission].

9.5.2 The Electrical Performance of WSi$_x$ in Devices

Good results have been reported [Saraswat et al.[225], Deal et al.[227], Trammel[228], Metz[228]] for CVD-WSi$_x$-polycide MOS devices. Even in the case where there was no poly-Si and thus the silicide was in direct contact with the gate oxide, very low levels of fixed oxide charge and interface traps were present. It is interesting how the workfunction found for polycide MOS

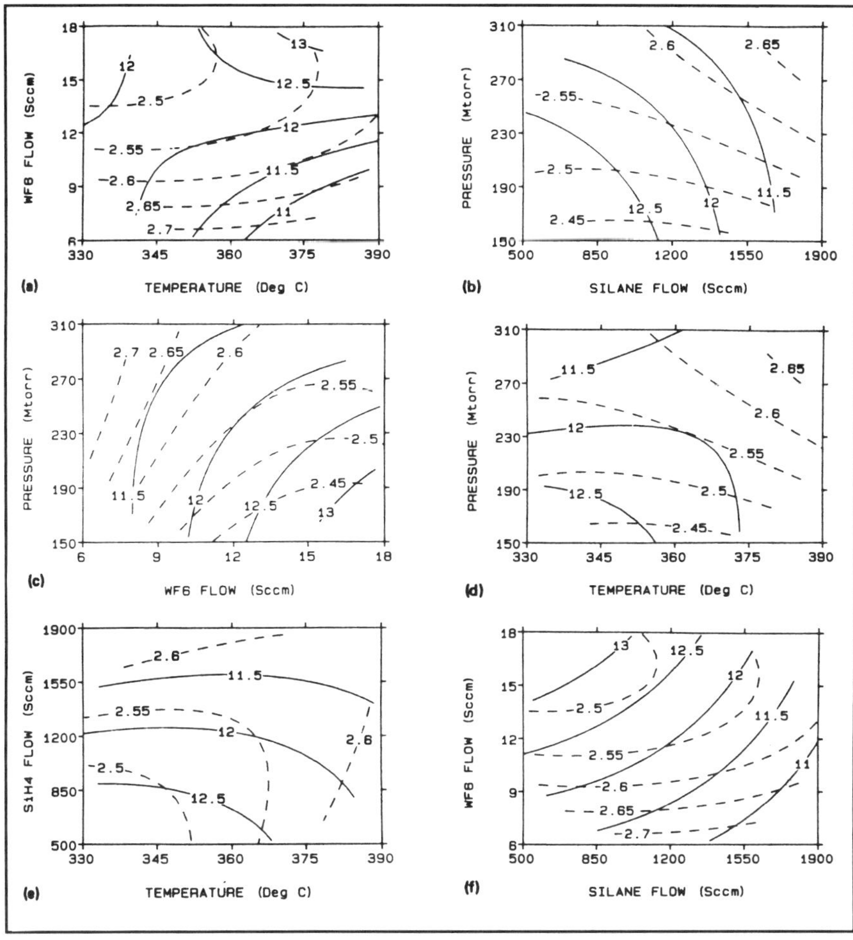

Figure 9.10. Contour plots of post annealed stress [10^9 dyne/cm^2 (-)] and as depos. Si/W ratio (--). Other variables are held at their center points. [Clark[226], reprinted with permission].

capacitors depends on the poly-Si thickness. For a poly-Si thickness of 100nm or less the workfunction of WSi$_2$ (4.9 eV) has been found. This in contrast to the capacitors with 250 or 500 nm of poly-Si, where the workfunction was that of phosphorous doped n$^+$ poly-Si (4.2 eV). Deal et al.[227], Trammel[228] and Metz[228] have found that existing poly-Si gate processes could be easy retrofitted by the polycide gate technology. No loss in yield or degradation in reliability could be associated with the use of the CVD-WSi$_x$.

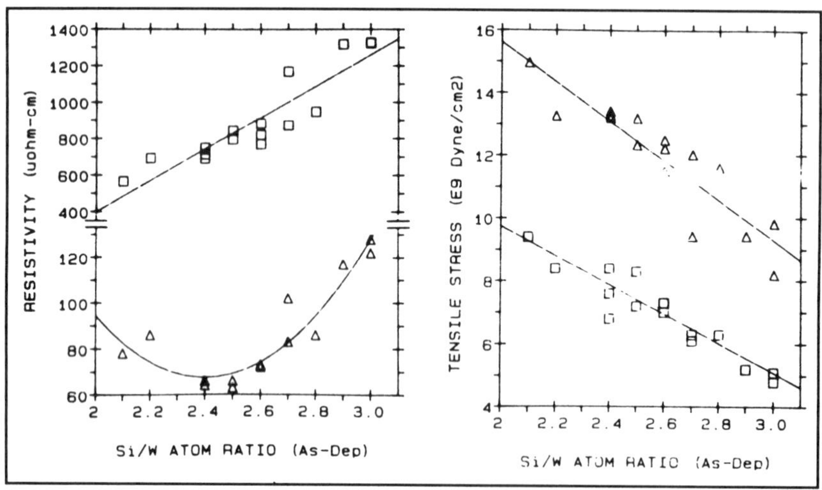

Figure 9.11. Resistivity and room temperature stress vs as- deposited stoichiometry before (□) and after anneal (▲). [Clark[226], reprinted with permission].

It should be noted that when gate oxides thicknesses become thinner and poly-Si line widths become narrower, certain problems emerge for SiH$_4$ based CVD-WSi$_x$. For more details see section 9.6.

WSi$_x$ has also been used as a direct wiring material between n$^+$-Si and p$^+$-Si. Excellent contact resistances were reported: 3×10^{-7} Ωcm^2 (n$^+$); 2×10^{-7} Ωcm$_2$ (p$^+$) and 1×10^{-7} Ωcm^2 for phosphorous doped poly-Si [Mihara et al.[215]].

9.5.3 Si Content in WSi$_x$

Saraswat et al.[225] noticed from an RBS analysis of as-deposited and annealed (1000°C, 10 min) CVD-WSi$_x$ films that starting from a Si/W atomic ratio of 2.5, this ratio decreased to 2.26 after anneal. They also showed a slight increase in the poly-Si thickness. A similar observation was made by Shioya et al.[229].

In a detailed study of polycide stack behavior during anneal, Kottke

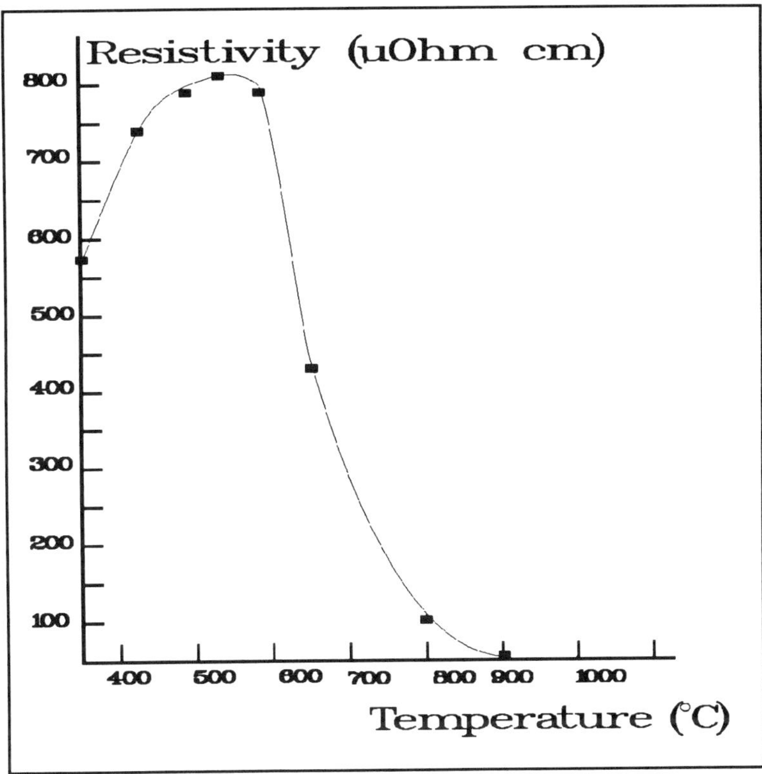

Figure 9.12. Resistivity of CVD-WSi$_x$ film as a function of the anneal temperature. As-deposited Si/W ratio is 2.3. [Data from ref. 232].

et al.[230] come to the following conclusions:

- Upon anneal there is a drop in the Si content of silicon rich silicide films. After a long enough anneal (two hours, 1000°C, N$_2$) all samples approach the 2.0 ratio but remain slightly Si rich. The resistivity of the films is then about 40 $\mu\Omega$cm. However, after only 30 min the resistivity already reaches 50 $\mu\Omega$cm.

- The excess Si precipitates to the poly-Si/WSi$_x$ interface. The Si removal from the silicide layer starts at this interface. As result, the thickness of the poly-Si increases and that of the WSi$_x$ film decreases.

- When the post-anneal resistivities are corrected for this film thickness shrink they become independent of the as-deposited Si/W atomic ratio. In figure 9.5, the as-deposited thicknesses were used for as-deposited and after anneal films to calculate the resistivities. Thus the increase in post-anneal bulk resistivity at higher Si/W ratios observed here is due to the thickness shrink during anneal.

The nature of the bonding between Si and W has been studied by Azizan et al.[216]. They found that the Si atoms had a charge loss of about 0.1-0.2 electron. This indicates that the bonding between Si and W has a strong covalent character. No indication can be found in the literature about the state of the excess Si in the as-deposited films. The question of whether it has a uniform bulk distribution or exists as a precipitate at the grain boundaries remains unanswered.

9.5.4 Thin Film Resistivity of WSi$_x$

An interesting discussion of the correlation between the resistivity of silicides and their electronic structure has been given by Sivaram et al.[265]. These authors show, for example, that because WSi$_x$ is a hole conductor (Malhotra et al.[265]), its resistivity will increase with increasing silicon content. The reader is referred to this article for a more detailed treatment.

The resistivity of WSi$_x$ as a function of anneal temperature exhibits a maximum value at about 550-600°C [LeGoues et al.[231], d'Heurle et al.[232], Shioya et al.[233]], as shown in figure 9.12. Such behavior has only been reported for WSi$_x$ and MoSi$_x$ films and not for the other refractory silicides. Although in practice anneal temperatures are always 900°C or above in order to achieve acceptable resistivities, it is interesting to consider what possible mechanisms cause this maximum value of the resistivity.

- Shioya et al.[233] determine the grain size using X-ray diffraction as a function of the anneal temperature. The grain size did not correlate with the resistivity of the film as one might expect on the ground of simple electron scatter theory (see also section 5.3). Also, no relationship could be detected between impurities like oxygen, carbon, fluorine or hydrogen and the resistivity. It was concluded that the high resistivity has to be attributed to the occurrence of the hexagonal phase which is observed in the X-ray diffraction pattern at about 600°C.

- LeGoues et al.[231] and d'Heurle et al.[232] pointed out that it is not likely that there will be a difference in resistivity between the hexagonal and tetragonal phase. In both phases the W atoms are coordinated to 10 silicon atoms and there is only a 2.5% difference in molar volume between the phases. The authors showed that the transition from the hexagonal phase to the tetragonal phase is accompanied by a maximum occurrence of stacking faults. These stacking faults will act as scattering centers thus accounting for the increase of the resistivity of the film at the phase transition. Their assumption is supported by TEM micrographs of the grains which show at 550°C 5×10^6 and at 800°C 2×10^4 stacking faults per cm. This is a difference of more than two orders of magnitude and could possibly account for the observed shift in resistivity.

9.5.5 Etching of WSi$_x$

Dry Etching: For reference purposes do we mention here briefly which chemistries have been reported to etch the WSi$_x$/poly-Si stack anisotropically, namely, CF_4/O_2, SF_6/O_2 and CCl_2F_2 [see Chow et al.[234] and references cited there, Chern et al.[234]]. With the appropriate reactor configuration and process conditions, good results are obtained. Etching can be done before or after anneal. The interested reader is referred to specialized literature for more details.

Wet Etching: One of the advantages of WSi$_2$ is that it is quite resistant to several chemicals in use in IC processing for cleaning or other purposes. Clark[226], investigated the resistance of WSi$_x$ to the following solutions: 10:1 HF, 7:1 BOE, hot (115°C) H_2SO_4, hot (85°C) HNO_3 and hot (115°C) H_2SO_4/H_2O_2. None of these solutions caused a change in sheet resistance of the exposed samples suggesting compatibility of the material with these solutions.

WSi$_2$ does etch in HF/HNO$_3$ solutions. However, if selectivity towards oxide or silicon is required, wet etching with this solution is not an appropriate candidate.

9.6 WSi$_x$ BASED ON SiH$_2$Cl$_2$/WF$_6$ CHEMISTRY

Silane based CVD-WSi$_2$ has been applied very successfully for years in several kinds of IC's. As mentioned earlier in this chapter the CVD deposition process has proven to be a good alternative for physical vapor deposition techniques. However, with the increasing integration and the accompanying decrease in gate width and gate oxide thickness, two problems become visible:

- The relatively high fluorine content (about 1 at%) in silane based CVD-WSi$_x$ process causes certain problems (see below) with thin (say <200Å) gate oxides.

- Cracking sometimes followed by delamination of the WSi$_x$ film is more pronounced with narrower poly-Si lines, higher steps and greater density of circuit components.

Studies have shown that WSi$_x$ based on SiH$_2$Cl$_2$/WF$_6$ chemistry can eliminate or suppress these problems. In the remainder of this section we will focus on the deposition process characterization of this chemistry.

The use of SiH$_2$Cl$_2$ as the reductant of WF$_6$ to obtain WSi$_x$ was first introduced by Price et al.[235] in 1986 at Spectrum CVD. The use of SiH$_2$Cl$_2$ instead of SiH$_4$ was originally justified as being a "cleaner" (particles) chemistry because of the lower reactivity of SiH$_2$Cl$_2$ compared to that of silane. This claim has not been proven yet, nor have particles been reported to be a problem of the silane chemistry. Nevertheless, soon after the fluorine content of the SiH$_2$Cl$_2$ process was shown to be three to four orders of magnitude lower than that of the SiH$_4$ process (see section 9.7). It was briefly pointed out by Price et al. that depending on the process conditions, Si/W atomic ratios from 1.3 to 2.7 in the temperature range 500-600°C

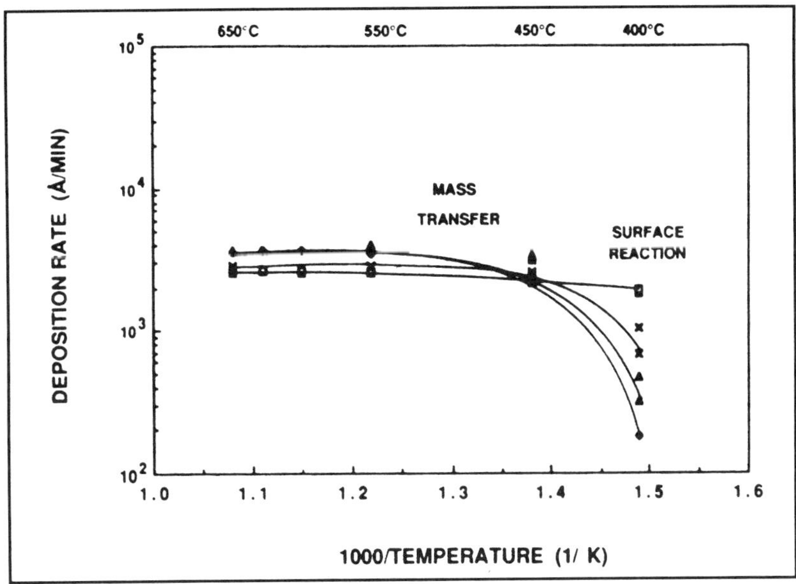

Figure 9.13. WSi$_x$ deposition rate vs. temperature. SiH$_2$Cl$_2$/WF$_6$ ratio constant at 32: □=64/2.0; x=80/2.5; ◊=128/4.0; ▲=176/5.5. [Tom Wu[236], reprinted with permission].

(wafer temperature) could be obtained.

A more detailed study was performed by Wu et al.[236] in 1988. The study was performed in a single wafer tool under conditions where there was a high reactant turnover (35%-60%). Therefore, from these data no conclusions about the true reaction kinetics can be drawn but valuable insight into the process parameters can be obtained. In figure 9.13, the deposition rate (using weight gain method) is plotted versus the reciprocal wafer temperature. Above a wafer temperature of 450°C deposition rate becomes independent of temperature. As has been pointed out by Srinivas et al.[237], this is not due to diffusion limitations but due to reactant starvation. Below 450°C, the reaction is surface rate controlled. The deposition rate dependence on SiH$_2$Cl$_2$ and WF$_6$ partial pressure is shown in figures 9.14 and 9.15. At 450°C, the deposition rate increases somewhat with increasing SiH$_2$Cl$_2$ partial pressure whereas at higher temperature there seems to be no dependence on the partial pressure. For WF$_6$ partial pressures less than 7 mTorr, however, the deposition rate does depend on

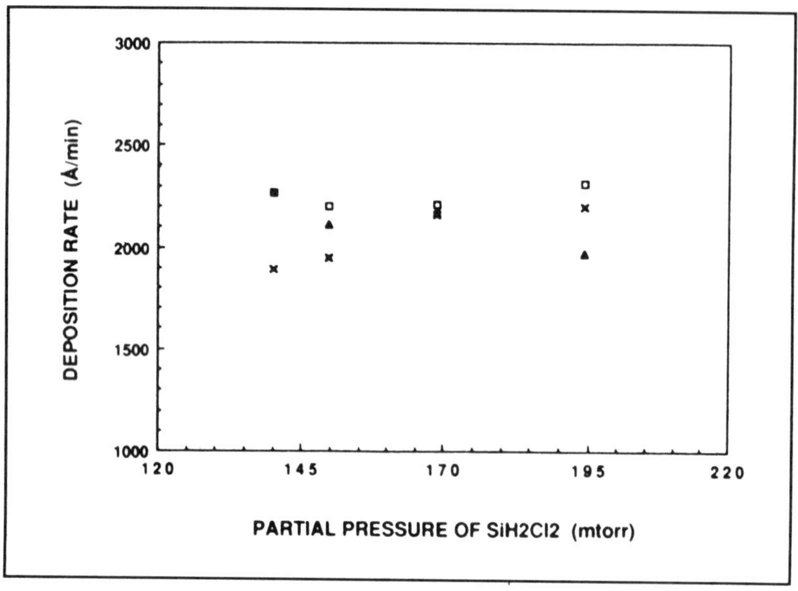

Figure 9.14. Deposition rate versus SiH$_2$Cl$_2$ partial pressure. WF$_6$ flow = 2.5 sccm, SiH$_2$Cl$_2$ flow = 60-100sccm, x=450°C; □=550°C; ▲=650°C.

the WF$_6$ flow. The as deposited resistivity exhibits an interesting behavior with respect to the temperature, at about 550°C where a maximum is observed (see figure 9.16). The reader will note that a similar behavior was observed for the low temperature anneal of silane based silicide films (see section 9.5.4). We refer to that section for a further discussion.

Selbrede[238], published a characterization of the SiH$_2$Cl$_2$-WSi$_x$ process in a batch reactor using an experimental design method. In comparing his results with those of Wu and Price we should keep in mind that the temperature in Selbrede's work are hot plate temperatures and that the actual wafer temperature can easily be 100°C lower in the pressure regime studied (see also chapter VII). The studied process window was:

Pressure:	300-600 mTorr
Temperature:	560-590°C
WF$_6$ flow:	10-16 sccm
SiH$_2$Cl$_2$ flow:	400-800 sccm.

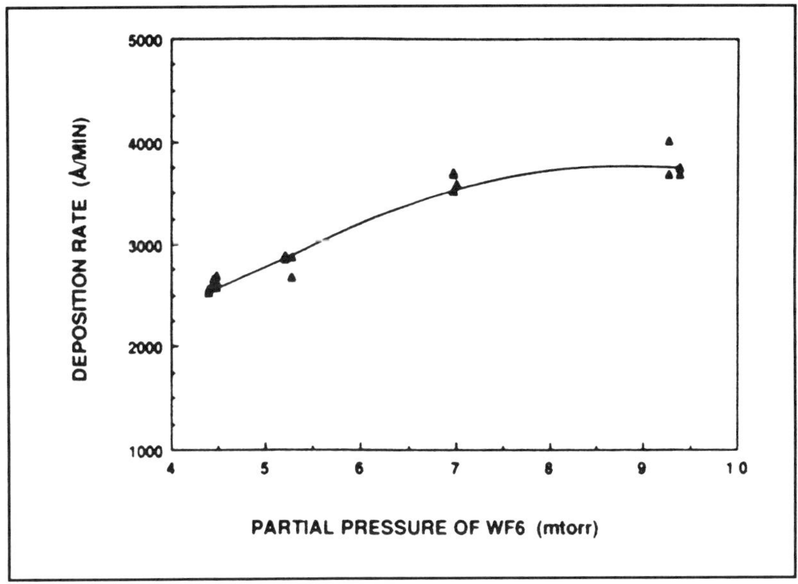

Figure 9.15. Deposition rate as a function of the WF$_6$ pressure. The SiH$_2$Cl$_2$/WF$_6$ ratio is held constant. [Tom Wu et al.[236], reprinted with permission].

Some results are shown in figure 9.17 and 9.18. It is important to note that the bulk resistivity for a given temperature can be adjusted by the SiH$_2$Cl$_2$ or the WF$_6$ flow to an acceptable value (normally close to 800 $\mu\Omega$cm). Also, there are clear differences between this work and that of Price and Wu (for instance discrepancies in the deposition rate dependence on the WF$_6$ and SiH$_2$Cl$_2$ flows). It is obvious that all studies were done while there were strong concentration gradients in the reactor. This condition makes partial pressure calculations uncertain and direct comparison between studies difficult. Nevertheless, process optimization is very well possible as shown by the study of Selbrede.

In an attempt to unravel the kinetics of the SiH$_2$Cl$_2$/WF$_6$ chemistry, Srinivas et al.[237], set up their experimental conditions such that the reactant conversion would be less than 10%. In this situation one can assume that the reactor is gradientless and inlet partial pressures (calculated from inlet flow ratios and the pressure) will be close to true wafer surface partial pressures. In calculating the conversion degree, however, one should keep

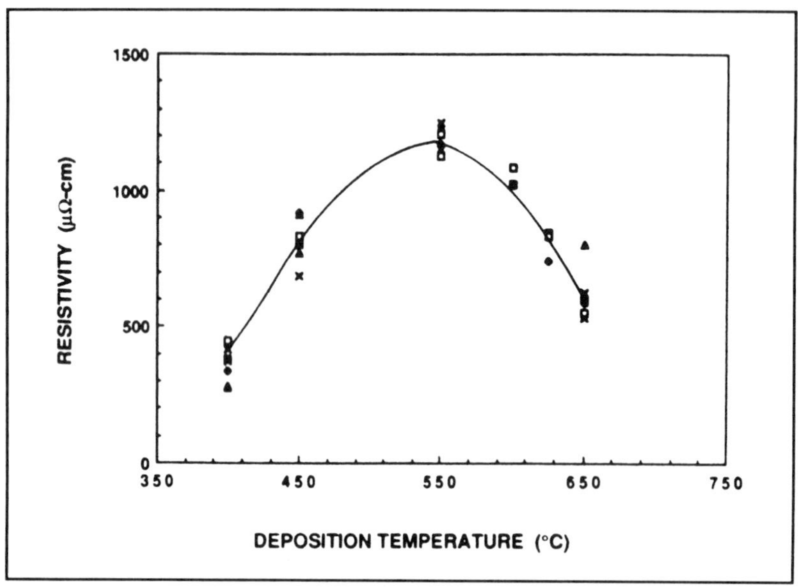

Figure 9.16. As-deposited resistivity as a function of the deposition temperature. SiH$_2$Cl$_2$/WF$_6$ ratio 32. [Tom Wu[236], reprinted with permission].

in mind that an appreciable amount (say 30%) of the reactants can still bypass the active reactor as was pointed out by Inamdar et al.[239]. A further complication is that the **composition** (ie. mainly the Si/W ratio) is a strong function of both the temperature and the gas phase composition as we saw above. Therefore, an Arrhenius plot to determine the activation energy or log(partial pressure) versus deposition rate plots to determine the reaction order cannot be made. Thus no kinetic conclusions can be drawn from deposition rate data gathered under these circumstances. For Si/W atomic ratios larger than 2 the following reactions will play a role:

$$SiH_2Cl_2 \ ---> \ Si + 2HCl \qquad (9.6)$$

and reaction 9.4:

$$2WF_6 + 10SiH_2Cl_2 \ ------> \ 2WSi_2 + 3SiF_4 + 3SiCl_4 + 8HCl + 6H_2$$

When we assign reaction rates R_{Si} and R_{WSi2} to reactions 9.6 and 9.4 respectively the as deposited Si/W atomic ratio can be expressed as:

196

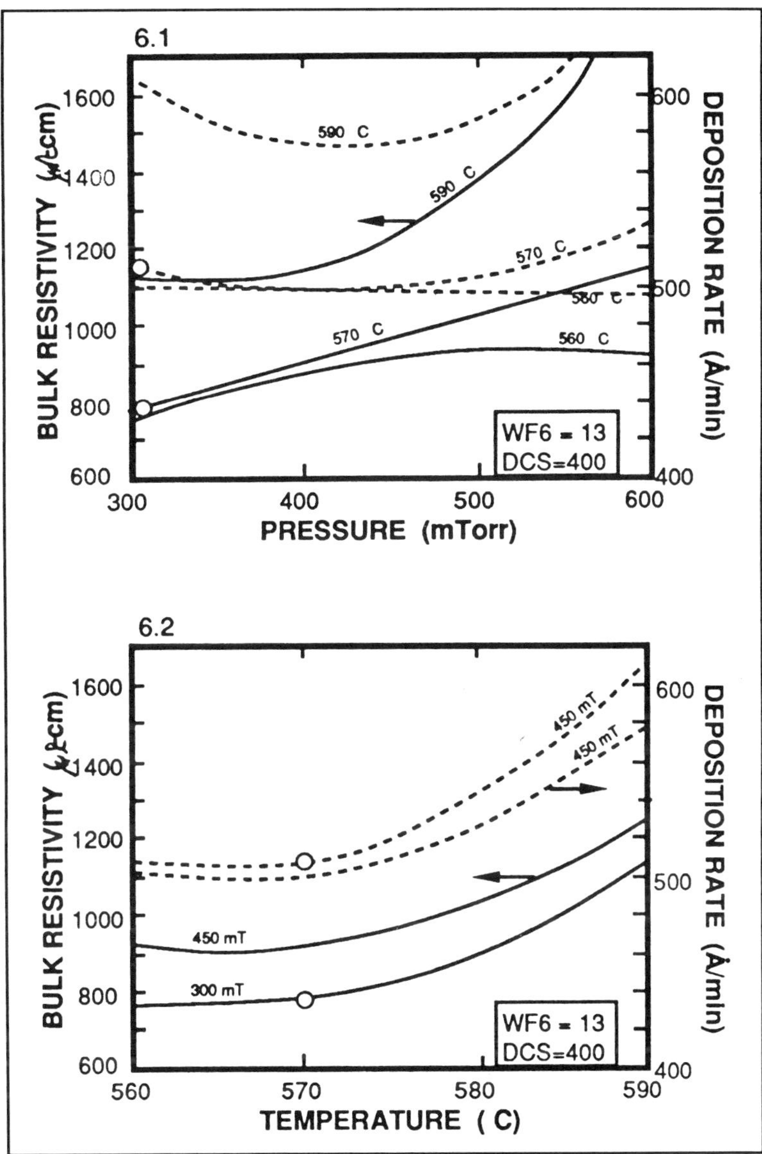

Figure 9.17. Thin film resistivity and deposition rate of SiH$_2$Cl$_2$-WSi$_x$ as a function of pressure and temperature. [Courtesy of S. Selbrede, Genus, Inc.].

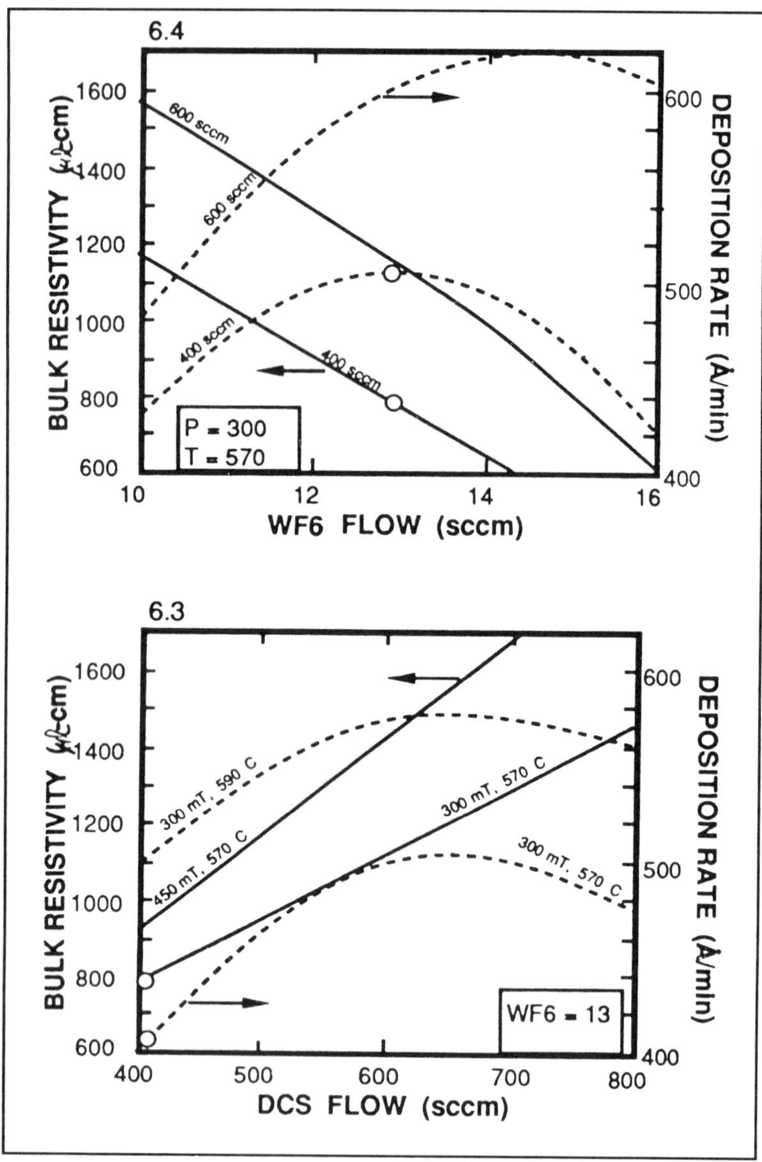

Figure 9.18. Thin film resistivity and deposition rate as a function of WF$_6$ and SiH$_2$Cl$_2$ flow. [Courtesy of S. Selbrede, Genus, Inc.].

$$Si/W = 2 + R_{Si}/R_{WSi2} \qquad (9.7)$$

Using absolute reaction rate theory for R_{Si} and R_{WSi2}, Srinivas et al. come to:

$$Si/W - 2 = k_{WSix}/k_{Si} \exp[-(E_{WSix}-E_{Si})/RT] \; x$$

$$x \; P_{SiH2Cl2}^{(nWSi2-nSi)} \; P_{WF6}^{(mWSi2-mSi)} \qquad (9.8)$$

where k_{WSi2} and k_{Si} are the rate constants; E_{WSi2} and E_{Si} are the activation energies; n_{WSi2} and n_{Si} are the reaction orders of SiH_2Cl_2 in reactions 9.4 and 9.6 and m_{WSi2} and m_{Si} are similar now for WF_6. Although WF_6 does not appear in the overall reaction 9.6, it may play a role in the decomposition of dichlorosilane at these low temperatures. Preliminary data of Srinivas et al. (at 460°C **wafer** temperature) shows that E_{WSi2}-E_{Si} = 70 kcal/mol and that n_{WSi2}-n_{Si}=+1 and m_{WSi2}-m_{Si}=-1. This equation gives a more quantitative description of the composition of the as deposited film in terms of temperature and reactant partial pressures (in a gradient less reactor!) and shows that both the SiH_2Cl_2 and WF_6 flows influence the film composition.

9.7 FLUORINE CONTENT IN CVD-WSi$_x$ FILMS

Evidence has been gathered that the fluorine content of SiH_4-WSi$_x$ films (typical number ~10^{20} at/cc) causes problems in the gate oxide [Fukumoto et al.[241], Shioya et al.[242], Wright et al.[243], Ellwanger et al.[244]]. Devices with gate oxides thinner than about 200Å exhibit specific problems (threshold voltage shift and low breakdown field) because of the fluorine incorporation by the gate oxide during anneal of the polycide stack. However, as we will see below, the accumulation of fluorine by the gate oxide also has positive effects (lower interface state concentration).

Fukumoto et al.[241] have found using SIMS analysis, that selective uptake of fluorine by the gate oxide after a 1000°C anneal of a CVD-WSi$_x$/poly-Si/SiO$_2$ stack occurs (see figure 9.19). In contrast, low fluorine

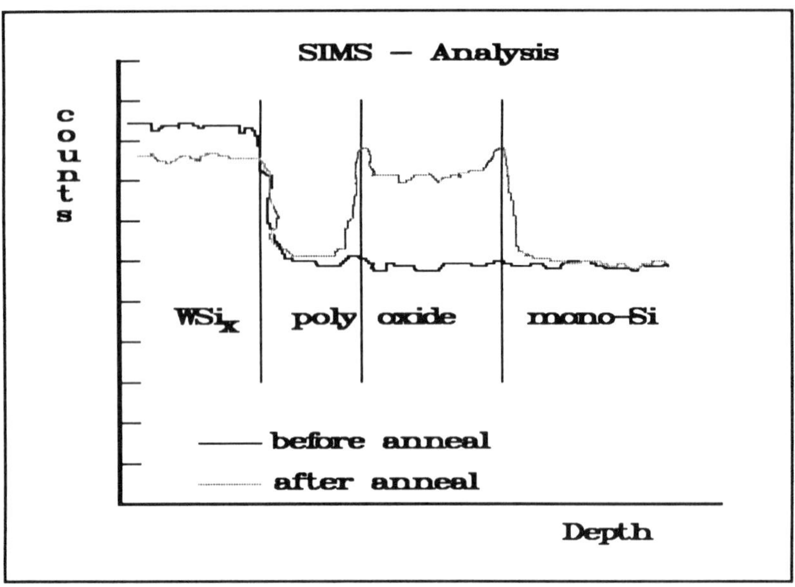

Figure 9.19 Schematic representation of a SIMS analysis of the fluorine profile in a polycide gate structure before and after anneal.

concentrations were present in the poly-Si layer and the mono-crystalline substrate. Their experiments suggest that the penetration of the fluorine through the poly-Si proceeds only by grain boundary diffusion. Therefore, the concentration of fluorine in the poly-Si remains low after anneal and no diffusion of fluorine into the mono-crystalline Si substrate could be observed.

Shioya et al.[242] gathered more evidence that indeed the fluorine diffusion (fluorine becomes mobile at about 800°C) into the gate oxide is causing the problem. They found that even after 1000°C anneal no hydrogen or W could be detected in the gate oxide using SIMS. The phosphorous in the poly-Si does diffuse into the WSi$_x$ film but not into the gate oxide.

A consistent model of the action of fluorine on gate oxide was proposed by Wright and Saraswat[243]. In this study the fluorine was introduced in the poly-Si by means of ion implantation. This approach allows various fluorine levels to be studied and eliminates complications from other elements like tungsten or hydrogen or stress induced effects. To

200

minimize effects due to implant damage, fluorine was implanted in the poly-Si layer. Neon implanted samples showed no degradation compared to unimplanted samples, indicating the validity of their experimental approach. Two gate oxide thicknesses were investigated: 13 and 41 nm. Important facts revealed from this study are:

- The diffusion of F occurs initially at 800°C. This confirms the result reached by Shioya et al.

- High frequency capacitance versus voltage measurements showed that for both oxide thicknesses there was a 10Å thickness increase at the highest implantation dose (10^{16}/cm^2). Since this can be caused by either a real thickness increase or by a decrease in dielectric constant, the actual thickness was verified using ellipsometry. The result was that there was a real thickness increase. More than a 200mV threshold voltage shift was observed!

- Only the highest implantation dose showed a minor degradation in the breakdown field, although, the charge to breakdown was severely degraded at the highest dose. Note that a dose of 10^{15}/cm^2 fluorine is approximately the same level as normally found in silane based WSi$_x$.

- Drift mobility experiments show that fluorine is not mobile in gate oxide, indicating that the fluorine is chemically bonded. XPS data shows no O-F bond (indicating that the fluorine is bonded to silicon atoms in the gate oxide).

- The concentration of the interface traps is the lowest for the highest implantation dose. This is in line with the result of Fukumoto et al. who found that the surface state concentration to be 1×10^{11}/cm^2 for sputtered WSi$_x$ but 7×10^9/cm^2 for CVD-WSi$_x$. Note that a typical surface state density for a good quality SiO$_2$/Si interface is in the order of 10^{10}/cm^2 or lower [Muller and Kamins[262]]. The conclusion is that fluorine deactivates the dangling bonds at the interface which are thought to be responsible for the surface states.

Wright and Saraswat[243] proposed a two step mechanism in which the fluorine bonds to dangling bonds at the SiO$_2$/Si interfaces and weakens bonds in the gate oxide. After saturation of the interface regions additional fluorine incorporation occurs in the bulk of the oxide by displacing oxygen. The liberated oxygen diffuses and forms new Si-O bonds explaining the increase in oxide thickness. Thus, while the interface regions are improved by the fluorine, excessive concentrations of fluorine degrade the bulk properties of the gate oxide as well as change the thickness of the gate oxide.

Ellwanger et al.[244] showed by using double poly-Si EPROM structures that in terms of gate oxide performance, low fluorine SiH$_2$Cl$_2$-WSi$_x$ (~ 10^{17} at/cc F) only marginally increased gate oxide thickness whereas SiH$_4$-WSi$_x$ (~ 10^{20} at/cc F) increased the thickness by about 20Å.

Table 9.2

Fluorine content of Silane-WSi$_x$ at different temperatures.

Temp. $^{\circ}$C	WF$_6$/SiH$_4$ sccm	resistivity $\mu\Omega$cm	fluorine at/cc
450	400/600	803	~ 10^{20}
500	400/600	1490	~ 10^{20}
610	400/550	4286	~ 4×10^{19}

From the discussion above it is clear that the high fluorine content of SiH$_4$ based CVD-WSi$_x$ might hamper its application in future IC's. It seems logical to investigate first if there exists possible process conditions where the film contains less fluorine. In the authors laboratories the following experiments were done (see table 9.2) in an attempt to lower the fluorine content. Higher deposition temperatures were investigated. Because the thin film resistivity increases sharply with the temperature, the WF$_6$/SiH$_4$ ratio had to be adjusted. We see that in the studied window no acceptable fluorine concentration can be achieved. Note, however, acceptable fluorine levels are not well established. From the work of Ellwanger et al. 1991 one can carefully deduce that the level should be below 10^{18} at/cc.

It has been well established that CVD-WSi$_x$ films based on SiH$_2$Cl$_2$/WF$_6$ chemistry have a fluorine level which can be as low as 2x10^{16} at/cc. The fluorine concentration is, within about one order of magnitude, insensitive for process parameters like temperature, pressure or reactant flows [Selbrede[238]] in a batch reactor. For a single wafer reactor the situation is far less clear as conflicting data exists [Price et al.[235], Wu et al.[236]]. The reason that the dichlorosilane chemistry produces lower fluorine levels remains unanswered although it has been suggested that it could be attributed to the higher deposition temperature as compared to the silane process [Selbrede[238]].

9.8 STRESS IN CVD-WSi$_x$ FILMS

The stress of annealed silicide films is tensile and is normally in the range 5-15x10^9 dyne/cm^2. In this section we will elaborate on how stress varies with the anneal temperature and with other process parameters. First, we will discuss some device problems which result from this stress, namely, crack formation and delamination. Cracking and delamination are here defined as follows:

- Cracking. When the tensile force surpass the cohesive forces inside the film, the film (line) breaks.

- Delamination. When the tensile force becomes larger than the adhesive force of the film towards the substrate and the film lifts. The adhesive properties of the film can be improved with a proper pretreatment of poly-Si prior to the silicide deposition. Both in situ (plasma cleaning [Nowicki et al.[250]] and stand alone cleaning [Ellwanger et al.[244]] have been shown to improve the adhesion of WSi$_x$ towards poly-Si.

Both cracking and delamination are the result of tensile stress and will reduce yield and/or reliability. Therefore, a few words about stress is appropriate.

Ellwanger et al.[244] compared the cracking and delamination behavior properties of SiH_4 and SiH_2Cl_2 silicide. In table 9.3 some results are gathered. From these data it is clear that a dichlorosilane film will usually exhibit superior delamination behavior, however, not necessarily always. The films with the highest Si/W ratio exhibit the best performance. Even for films with identical Si content (ranking 4 and 5) the SiH_2Cl_2-WSi$_x$ film out performs the SiH_4-WSi$_x$ film. We will discuss a possible reason for this below. The situation is dramatically illustrated in figure 9.20.

The improved mechanical properties (such as stress, adhesion and oxidation stability) with increasing silicon content have been noticed in other studies. Brors et al.[217], have reported that the Si/W ratio should stay above 2.0 in order to have stable films. Shishino et al.[245], in a study of open failures in WSi$_x$ interconnects, come to the conclusion that by increasing the Si/W ratio from 2.44 to 2.58, the failure rate is decreased from 0.350% to 0.007%. Shioya et al.[246] have shown that the room temperature stress reduces sharply with increasing silicon content for silane based films. For a SiH_2Cl_2 based film the dependence of the film stress on process parameters is less clear cut [Selbrede[238]].

Table 9.3

Percentage of WSi$_x$ delamination on various widths of poly-Si runners

---

Chemistry	Si/W*	0.8μm	0.6μm	0.4μm	Rank
SiH_2Cl_2	3.5	0%	0%	4%	1
"	3.3	0	0	22	2
"	2.9	0	0	99	3
"	2.6	0	19	100	4
SiH_4	2.6	1	70	100	5
SiH_2Cl_2	2.3	100	100	100	6

Data from Ellwanger et al.[244], with permission.

* as-deposited film, silicon over tungsten ratio.

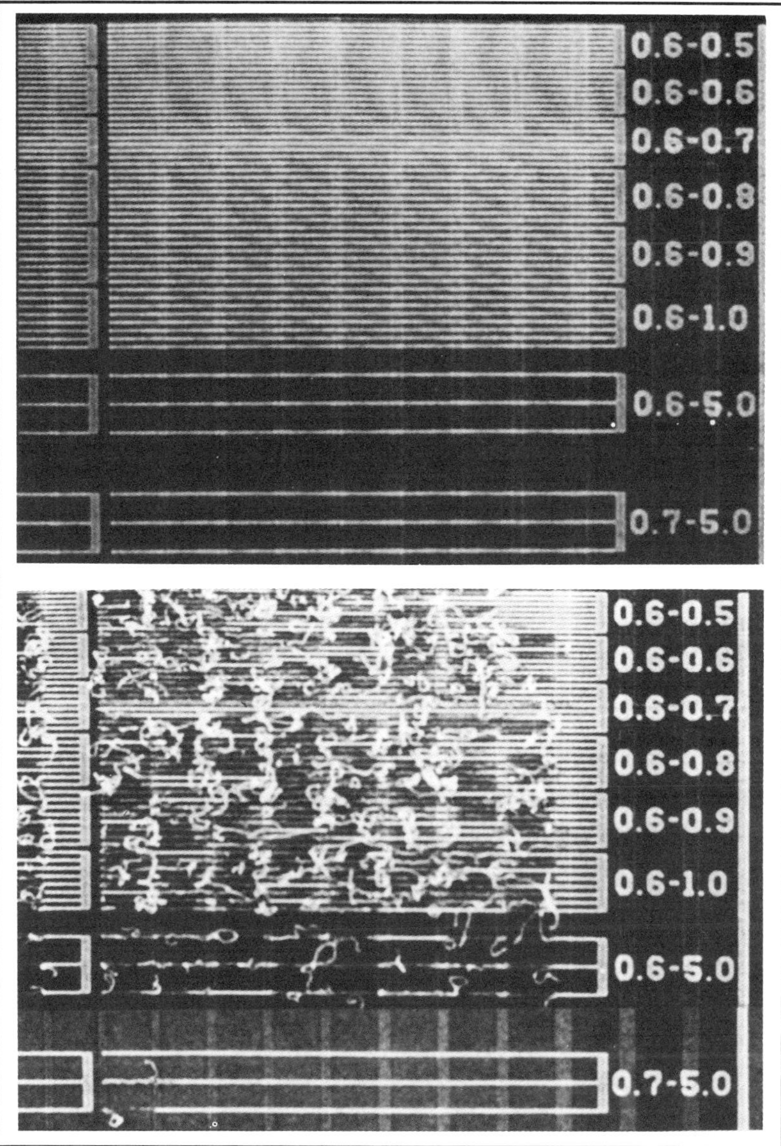

Figure 9.20. WSi$_x$ delamination of 0.6 and 0.7μm wide polycide lines. Top SiH$_2$Cl$_2$ (x=2.9); SiH$_4$ (x=2.6) bottom. [Ellwanger et al.[244], reprinted with permission].

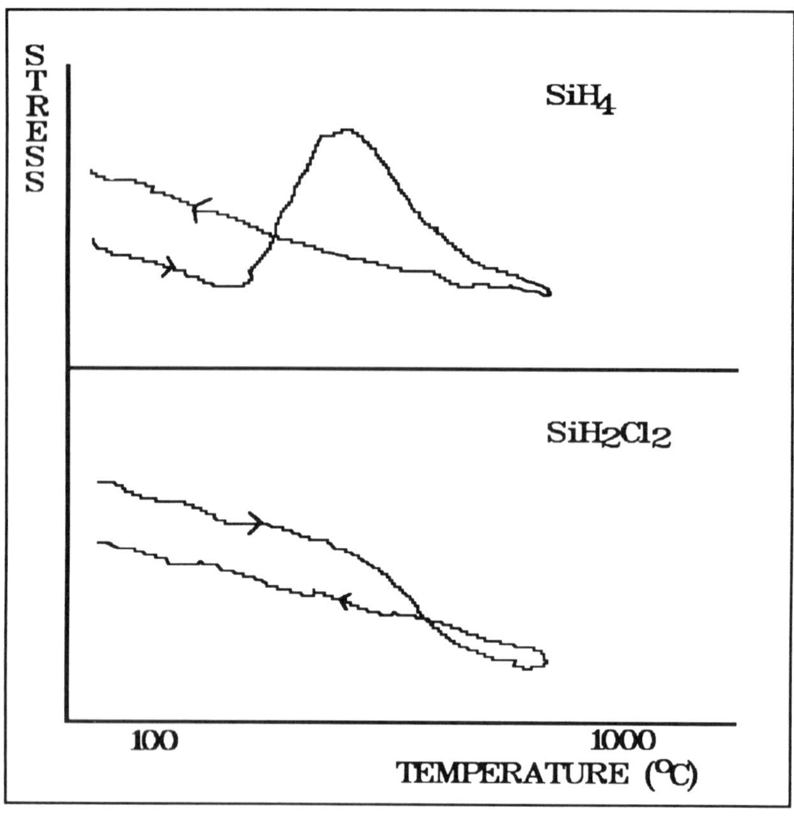

Figure 9.21. Schematic representation of stress behavior of SiH$_4$-WSi$_x$ (top) and SiH$_2$Cl$_2$-WSi$_x$ (bottom) films during anneal.

As can be seen from table 9.3, in the case of identical Si/W ratio (2.6), the dichlorosilane based film ranks higher than the silane based film in terms of delamination behavior. A possible explanation for this might be the stress development upon annealing of the film [Shioya et al.[247], Hara et al.[247]] as demonstrated in figure 9.21. The silane based film exhibits a sharp increase in stress at about 350°C. This has been attributed to crystallization of the film (amorphous → hexagonal → tetragonal). Note that the room temperature stress after anneal is **higher** than before anneal. In the dichlorosilane case there is no maximum in the stress profile and moreover, the stress after anneal is **lower** than the stress before anneal. This lack of a maximum stress during anneal may be attributed to the higher deposition temperature of the dichlorosilane WSi$_x$ film (>600°C versus 360°C chuck

206

temperature). The higher deposition temperature causes in the dichlorosilane silicide crystallization during deposition whereas the silane film is completely amorphous in the as-deposited state [Hara et al.[240]]. Thus the variation in stress during an anneal cycle is much larger in the SiH$_4$-WSi$_x$ film than in the SiH$_2$Cl$_2$-WSi$_x$ film. This might very well be the reason for the improved mechanical performance of the latter film.

9.9 STEP COVERAGE OF CVD-WSi$_x$ FILMS

Intuitively one can imagine that the step coverage of the silicide layer is an important parameter in the degree of formation of cracks and subsequent delamination. Undoubtedly, the step coverage of evaporation or sputter deposition techniques can be extremely poor, and in addition the step coverage may vary from wafer to wafer depending on the location of the wafer in the chamber during deposition. Thus delamination or cracking problems can be expected with these deposition techniques.

Although the step coverage of CVD-WSi$_x$ when compared with, for instance, that of CVD-W (using the H$_2$/WF$_6$ chemistry, see chapter III) is probably poor, it is superior to that of the PVD methods. Rode et al.[223] showed the step coverage of CVD-WSi$_x$ over an almost vertical step to be better than 80%. It has been claimed that the step coverage of the SiH$_2$Cl$_2$/WF$_6$ WSi$_x$ chemistry may be better than that of the SiH$_4$/WF$_6$ WSi$_x$ chemistry [Rode et al.[223], Selbrede[238]]. More evidence has been provided recently by Hillman et al.[248].

Raupp et al.[249] have modelled the step coverage of the SiH$_2$Cl$_2$ silicide system following basically the same approach as used in the step coverage discussion in chapter II. From their model, the conclusion can be drawn that with increasing temperature and increasing reactant turnover (or starvation of the reactor) the step coverage will degrade. The observed experimental trend followed indeed these predictions.

9.10 CONCLUSIONS

The discussions in this chapter show clearly that:

- CVD as a deposition method for WSi$_x$ has superior properties over sputtering or evaporation methods.

- Although the electrical performance of silane based CVD-WSi$_x$ in devices is extremely good and can be easily retrofitted into existing poly-Si gates, the high fluorine content, its moderate step coverage and particular stress behavior can limit its use in future generation IC's.

- CVD-WSi$_x$ based on dichlorosilane looks very promising and appears to solve the problems associated with silane based tungsten silicide.

REFERENCES

The references are grouped according to their main subject. Additional material can be found in the Proceedings of the Workshop on Tungsten and Other Refractory Metals for VLSI Applications (I-VI) published by the Materials Research Society. Another valuable literature source especially from the point of integration are the Proceedings of the IEEE International VLSI Multilevel Interconnection Conference (VMIC).

The following abreviations are used:

- Tungsten Workshop I = Proceedings of the Workshop on "Tungsten and Other Refractory Metals for VLSI Applications", R.S. Blewer ed., Materials Research Society, Pittsburgh PA.

- Tungsten Workshop II = Proceedings of the Workshop on "Tungsten and Other Refractory Metals for VLSI Applications II", E.K. Broadbent ed., Materials Research Society, Pittsburgh PA.

- Tungsten Workshop III = Proceedings of the Workshop on "Tungsten and Other Refractory Metals for VLSI Applications III", V.A. Wells ed., Materials Research Society, Pittsburgh PA.

- Tungsten Workshop IV = Proceedings of the Workshop on "Tungsten and Other Refractory Metals for VLSI Applications IV", R.S. Blewer and C.M. McConica ed., Materials Research Society, Pittsburgh PA.

- Tungsten Workshop V = Proceedings of the Workshop on "Tungsten and Other Advanced Metals for VLSI/ULSI Applications V", S.S. Wong and S. Furukawa ed., Materials Research Society, Pittsburgh PA.

- Tungsten Workshop VI = Proceedings of the Workshop on "Tungsten and Other Advanced Metals for VLSI/ULSI Applications 1990", G.C. Smith and R. Blumenthal ed., Materials Research Society, Pittsburgh PA.

References

Adhesion layers.
[1] V. Hoffman, Solid State Techn, june, 119 (1983).
[2] S.S. Cohen, M.J. Kim, B. Gorowitz, R. Saia, T.F. McNelly and G.Todd, Appl. Phys. Lett., **45**(4), 414 (1984).
[3] A. Kohlhase, M. Mändl and W. Pamler, J. Appl. Phys., **65**(6), 2464 (1989).
[4] R.C. Ellwanger and J.M. Towner, Thin Solid Films, **161**, 289 (1988).
[5] M. Wittmer, J. Vac. Sci. Technol., **A3**(4), 1797 (1985).
[6] S.E. Babcock and K.N. Tu, J. Appl. Phys., **59**(5), 1599 (1986).
[7] R.C. Ellwanger, J.E.J. Schmitz, R.A.M. Wolters and A.J.M. van Dijk, Tungsten Workshop II, 385 (1987).
[8] V.V.S. Rana, J.A. Taylor, L.H. Holschwandner and N.S. Tsai, Workshop II, 187 (1987).
[9] M. Iwasaki, H. Itoh, T. Katayama, K. Tsukamoto and Y. Akasaka, Tungsten Workshop V, 187 (1990).
[10] K.C. Ray Chiu and N.E. Zetterquist, Tungsten Workshop II, 177 (1987).

Encroachment and wormholes.
[11] W.T. Stacy, E.K. Broadbent and M.H. Norcott, J. Electrochem. Soc., **132**(2), 444 (1985).
[12] E.K. Broadbent, A.E. Morgan, J.M. DeBlasi, P. van der Putte, B. Coulman, B.J. Burrow and D.K. Sadana, J. Electrochem. Soc., **133**(8), 1716 (1986).
[13] P. van der Putte, D.K. Sadana, E.K. Broadbent and A.E. Morgan, Appl. Phys. Lett., **49**(25), 1723 (1986).
[14] R.C. Ellwanger, A.J.M. van Dijk, J.E.J. Schmitz and R.D.J. Verhaar, Tungsten Workshop, 93 (1989).

CVD of TiN.
[15] S.R. Kurtz and R.G. Gordon, Thin Solid Films, **140**, 277 (1986).
[16] N. Yokoyama, K. Hinode and Y. Homma, J. Electrochem. Soc, **136**(3), 882 (1989).
[17] F. Pintchovski, T. White, E. Travis, P.J. Tobin and J.B. Price, Tungsten Workshop IV, 275 (1989).
[18] A. Sherman, J. Electrochem. Soc., **137**(6), 1892 (1990).
[19] K. Ikeda, M. Maeda, Y. Arita, IEEE, Symp. on VLSI Techn., 61 (1990).

[20] M.J. Buiting, A.F. Otterloo and A.H. Montree, J. Electrochem. Soc., **136**(2), 500 (1991).

[21] N. Nakanishi, S. Mori and E. Kato, J. Electrochem. Soc, **137**(1), 322 (1990).

[22] G.C. Smith and D. Yin, Workshop VI, 267 (1991).

[23] I.J. Raaijmakers and A. Sherman, Proc. 7th Int. IEEE VLSI Multilevel Interconnection Conference, 219 (1990).

Contact resistance measurement.

[24] B. Pellegrini and G. Salardi, Solid State Electronics, **18**, 791 (1975).

[25] A.A. Naem and D.A. Smith, J. Electrochem. Soc., **133**(11), 2377 (1986).

[26] W.M. Loh, S.E. Swirhun, T.A. Schreyer, R.M. Swanson and K.C. Saraswat, IEEE Trans. E.D., **ED-34**(3), 512 (1987).

[27] A. Scorzoni and M. Finetti, Material Science Reports, **3**, 79 (1988).

[28] P. J. Wright, W.M. Loh and K.C. Saraswat, IEEE Trans. E.D., **35**(8), 1328 (1988).

Surface reactions.

[29] M.L. Yu and B.N. Eldridge, J. Vac. Sci. Technol., **A7**(3), 625 (1989).

[30] S.Sivaram, E. Rode and R. Shukla, Tungsten Workshop V, 47 (1990).

Step coverage.

[31] J.E.J. Schmitz, R.C. Ellwanger and A.J.M. van Dijk, Tungsten Workshop III, 55 (1988); R. Blumenthal and G.C. Smith, ibid, p.47

[32] A. Hasper, J. Holleman, J. Middelhoek and C.R. Kleijn, Tungsten Workshop V, 127 (1990).

[33] R.M. Levin and K. Evans-Lutterodt, J. Vac. Sci. Technol., **B1**(1), 54 (1983).

[34] C.C. Tsai, J.C. Knights, G. Chang and B. Wacker, J. Appl. Phys., **59**(8), 2998 (1986).

[35] J.G. Shaw and C.C. Tsai, J. Appl. Phys., **64**(2), 699 (1988).

[36] A. Yuuki, Y. Matsui and K. Tachibana, Jap. J. Appl. Phys., **28**(2), 212 (1989).

[37] L.Y. Cheng, J.P. McVittie and K.C. Saraswat, Proc. Electrochem. Spring Meet., Ext. Abstr. Vol. 89-1, 263 (1989).

[38] Y. Okada, J. Chen, I.H. Campbell, P.M. Fauchet and S. Wagner, J. Appl. Phys., **67**(4), 1757 (1990).

[39] C.M. McConica and S. Churchill, Tungsten Workshop III, 257 (1988).

[40] S. Chatterjee and C.M. McConica, J. Electrochem. Soc, **137**(1), 328 (1990).

[41] J.R. Dacey, Industrial and Engineering Chemistry, **57**(6), 27 (1965).

[42] R.M. Barrer, Appl. Material Res., **2**(3), 129 (1963).

[43] J.E.J. Schmitz, W.L.N. van der Sluys and A.H. Montree, Tungsten Workshop V, 117 (1990).

Kinetics of CVD-W.

H_2/WF_6 chemistry.

[44] E.K. Broadbent and C.L. Ramiller, J. Electrochem. Soc., **131**(6), 1427 (1984).

[45] Y. Pauleau and Ph. Lami, J. Electrochem. Soc., **132**(11), 2779 (1985).

[46] C.M. McConica and K. Krishnamani, J. Electrochem. Soc., **133**(12), 2542 (1986).

[47] H. Cheung, in Proc. 3rd Int. Conf. on CVD, F.A. Glaski, The Am. Nucl. Soc., Hinsdale Il., p. 136 (1972).

SiH_4/WF_6 chemistry.

[48] J.E.J. Schmitz, A. van Dijk and M. Graef, Proc. 10th Int. Conf. on CVD, The Electrochem. Soc., Vol. 87-8, 625 (1987).

[49] R.S. Rosler, J. Mendonca and M.J. Rice Jr., J. Vac. Sci. Technol. **B6**(6), 1721 (1988).

High pressure blanket tungsten.

[50] J.E.J. Schmitz, A.J.M. van Dijk, J.L.G. Suijker, M.J. Buiting and R.C. Ellwanger, Appl. Surface Science, **38**, 350 (1989).

[51] R.V. Joshi, E. Mehter, M. Chow, M. Ishaq, S. Kang, P. Geraghty and J. McInerney, Tungsten Workshop V, 157 (1990).

[52] T.E. Clark, A.P. Constant, M. Chang and C. Leung, Tungsten Workshop V, 167 (1990); see also T.E. Clark, Microelectronic Manufacturing and Testing, june, 27 (1990).

[53] E.J. McInerney, P. Geraghty and S. Kang, Tungsten Workshop V, 135 (1990).
[54] L. Bartholomew and G. Max McDaniel, Tungsten Workshop VI, 55 (1991).

Etching of tungsten.

[55] M.E. Burba, E. Degenkolb, S. Henck, M. Tabasky, E.D. Jungbluth and R. Wilson, J. Electrochem. Soc., **133**(10), 2113 (1986).
[56] D.W. Hess, Solid State Technol., april, 97 (1988).
[57] T.H. Daubenspeck and P.C. Sukanek, J. Electrochem. Soc, **136**(12), 3779 (1989).
[58] N. Mutsukura and G. Turban, J. Electrochem. Soc., **137**(1), 225 (1990).
[59] P.E. Riley, M. Chang, S.G. Ghanayem and A. Mak, IEEE Trans. Semicond. Manufact., **3**(3), 142 (1990).
[60] R. Nowicki, S. Otto and P. Geraghty, to be published.
[61] J.M.F.G. van Laarhoven, H.J.W. van Houtum and L. de Bruin, Proc. 6th Int. VLSI Multilevel Interconn. Conf., IEEE, 129 (1989).
[62] J. Berthold and C. Wieczorek, Applied Surface Science, **38**, 506 (1989).
[63] G. Higelin, C. Wieczorek and V. Grewal, Proc of the 3rd Int. VLSI Multilevel Inrtconn. Conf. IEEE, 29 (1986).
[64] G.C. Smith and R.B. Jucha, Proc. of the V-MIC Conf, 403 (1986).
[65] G. de Graaf, A.L. Butler and R. Penning de Vries, Proc. of the 3rd Int. VLSI Multilevel Conf., IEEE, 357 (1988).

Selective tungsten.

Si/WF_6 and H_2/WF_6 chemistry.
[66] R.H. Wilson and R.W. Stoll, Tungsten Workshop III, 311 (1988).
[67] R. Chow, J. Schmitz, P. Arnold, J.T. Gasner and J.D. Butler, Tungsten Workshop VI, 89 (1991).
[68] T. Ohba, S. Inoue and M. Maeda, IEDM, 214 (1987).
[69] T. Ohba, T. Suzuki and T. Hara, Tungsten Workshop IV, 17 (1989).
[70] S.R. Herd, K.Y. Ahn, P.M. Fryer and J.M. Karasinski, Tungsten Workshop IV, 47 (1989).

[71] K.Y. Ahn, T. Lin and J. Angilello, Tungsten Workshop III, 25 (1988).

[72] A.E.T. Kuiper, M.F.C. Willemsen and J.E.J. Schmitz, Applied Surface Science, **38**, 338 (1989).

[73] M.L. Green, Y.S. Ali, T. Boone, B.A. Davidson, S.C. Feldman and S. Nakahara, J. Electrochem. Soc., **134**, 2285 (1987).

[74] M.L. Hitchman, A.D. Jobson and L.F.Tz. Kwakman, Applied Surface Science, **38**, 312 (1989); for more details see [73], [107], [120] and

K.Y. Tsao and H.H. Busta, J. Electrochem. Soc., **131**(11), 2702 (1984)

M.L. Green and R.A. Levy, ibid, **132**(5), 1243 (1985)

M.Wong, N. Kobayashi, R. Browning, D. Paine and K.C. Saraswat, ibid, **134**(9), 2339 (1987).

[75] M.L. Yu, B.N. Eldridge and R.V. Joshi, Tungsten Workshop III, 75 (1988).

[76] R.A. Levy, M.L. Green, P.K. Gallagher and Y.S. Ali, J. Electrochem. Soc., **133**(9), 1905 (1986).

[77] R.S. Blewer, T.J. Headley and M.E. Tracey, Tungsten Workshop III, 115 (1988); J.M. DeBlasi, D.K. Sadana and M.H. Norcott, Mat. Res. Symp. Proc. Vol.71, 303 (1986).

[78] A. Hårsta and J.O. Carlsson, Tungsten workshop V, 77 (1990).

[79] M.L. Yu, K.Y. Ahn and R.V. Joshi, Tungsten Workshop V, 15 (1990).

[80] C. Fuhs, E.J. McInerney, L. Watson and N. Zetterquist, Tungsten Workshop I, 257 (1985).

SiH_4/WF_6 chemistry.

[81] R.F. Foster, S. Tseng, L. Lane and K.Y. Ahn, Tungsten Workshop III, 69 (1988).

[82] Y. Kusumoto, K. Takakuwa, H. Hashinokuchi, T. Ikuta and I. Nakayama, Tungsten Workshop III, 103 (1988).

[83] J.E.J. Schmitz, M.J. Buiting and R.C. Ellwanger, Tungsten Workshop IV, 27 (1989).

[84] K. Kajiyana, K. Tsunenari and T. Kikkawa, Tungsten Workshop V, 39 (1990).

[85] H. Itoh, R. Nakata, N. Kaji, T. Endo, T. Watanabe and H. Okano, Mat. Res. Symp. Proc. VLSI V, p. 23 (1990).

References

Substrate materials.
[86] S.P. Murarka, "Silicides for VLSI Applications", Ac. Press, New York, (1983).
[87] L. Van den Hove, K. Maex, L. Hobbs, P. Lippens, R. De Keersmaecker, V. Probst and H. Schaber, Applied Surface Science, **38**, 430 (1989).
[88] R.D.J. Verhaar, A.A. Bos, J.M.F.G. van Laarhoven, H. Kraaij and R.A.M. Wolters, Applied Surface Science, **38**, 458 (1989).
[89] R.Wolters and L. Van den Hove, Proc. 3rd Int. VLSI Multilevel Conf, IEEE, 149 (1988).
[90] T.E. Tang, C. Wei, R.A. Haken, T.C. Holloway, L.R. Hite and T.C.W. Blake, IEEE Trans Electron Dev., **ED-34**(3), 682 (1987).
[91] F.M. D'Heurle, Metall. Trans.,**1**, 683 (1971).
[92] N. Hirashita, M. Kinoshita and T. Ajioka, J. Electrochem. Soc., **135**(12), 3159 (1988).
[93] A.J. Learn, J. Electron. Materials, **3**(2) (1974).
[94] H. Oikawa and T. Amazawa, ECS 3rd Int. Symp. VLSI Sci and Technol., **85-5**, 131 (1985).
[95] C. Ting and M. Paunovic, J. Electrochem. Soc., **136**, 456 (1989).
[96] K. Haberle, W. Langheinrich, V. Dudek, H.A. Hefner and R. Isernhagen, Proc. 5th Int. VLSI Multilevel Interconn. Conf., IEEE, 117 (1988).
[97] D. Yen and G. Rao, ibid, p85 (1988).
[98] S.Chen, Y. Chao, J.J. LIn and F.C. Tseng, ibid, 306 (1988).
[99] M. Delfino and D.HG. Choe, Tungsten Workshop IV, 57 (1989).
[100] R.W. Pattee, C.M. McConica and K. Baughman, J. Electrochem. Soc., **135**(6), 1477 (1988).

Electrical contact characterisation
[101] K.C. Saraswat, S. Swirhun and J.P. McVittie, Proc. Electrochem. Soc., **84-7**, 409 (1984).
[102] T. Tsutsumi, H. Kotani, J. Komori and S. Nagao, IEEE Trans. Electr. Dev., **37**(3), 569 (1990).
[103] S.S. Cohen, J. Appl. Phys., **59**(6), 2072 (1986).
[104] H. Itoh, T. Moriya and M. Kashiwagi, Solid State Technol. nov, 83 (1987).
[105] T.V. Nordstrom and J.P. Whitlock, Tungsten Workshop IV, 159 (1989).

[106] F. Matsuoka, H. Iwai, K. Hama, H. Itoh, R. Nakata, T. Nakakubo, K. Maeguchi and K. Kanzaki, IEEE Trans. Electr. Dev., **37**(3), 562 (1990).

Barrier properties CVD-W.
[107] R.V. Joshi and D.A. Smith, Mat. Res. Soc. Symp. Proc. Vol **71**, 309 (1986).
[108] Y. Shioya, M. Maeda and K. Yanagida, J. Vac. Sci. Technol. B, **4**(5), 1175 (1986).
[109] Y. Pauleau, Ph. Lami, A. Tissier, R. Pantel and J.C. Oberlin, Thin Solid Films, **143**, 259 (1986).
[110] O. Thomas, A. Charai, F.M. D'Heurle, T.G. Finstad and R.V. Joshi, Thin Solid Films, **171**, 343 (1989).

Selective tungsten on silicide.
[111] E.K. Broadbent, A.E. Morgan, J.M. DeBlasi, P. van der Putte, B. Coulman, B.J. Burrow, D.K. Sadana and A. Reader, J. Electrochem. Soc., **133**(8), 1715 (1986).
[112] P. van der Putte, D.K. Sadana, E.K. Broadbent and A.E. Morgan, Appl. Phys. Lett., **49**(25), 1723 (1986).
[113] R.C. Ellwanger, J.E.J. Schmitz and A.J.M. van Dijk, Tungsten Workshop III, 399 (1988).
[114] G.C. Smith and R.B. Jucha, Proc. 3rd Int. VLSI Multilevel Interconn. Conf., IEEE, 403 (1986).
[115] S.L. Ng, S.J. Rosner, S.S. Laderman, T.I. Kamins, D.R. Breadbury and J. Amano, Tungsten Workshop II, 93 (1987).

β-Tungsten.

[116] W.R. Morcom, W.L. Worrell, H.G. Sell and H.I. Kaplan, Metallurgical Trans., **5**, 155 (1974).
[117] C.C Tang and D.W. Hess, Appl. Phys. Lett., **45**(6), 633 (1984).
[118] D.C. Paine, J.C. Bravman, C.Y. Yang, Appl. Phys. Lett., **50**(9), 498 (1987).
[119] D. Davazoglou and A. Donnadieu, Thin Solid Films, **147**, 131 (1987).
[120] H.H. Busta and C.H. Tang, J. Electrochem. Soc., **133**, 1195 (1986).

[121] G. Hagg and N. Schonberg, Acta Crystallogr., 7, 351 (1954).

Selectivity loss.

[122] R.S. Blewer, Solid State Technol., 178 (1986).
[123] Ph. Lami and Y. Pauleau, J. Electrochem. Soc., **135**, 980 (1988).
[124] L.F.Tz. Kwakman, W.J.C. Vermeulen, E.H.A. Granneman and M.L. Hitchman, Tungsten Workshop III, 141 (1988).
[125] C. McConica, Tungsten Workshop II, 51 (1987).
[126] S. Tooru Sumiya, I. Hirase, D. Rufin, S. Ukishima, M. Schack, M. Shishikura, M. Matsuura and A. Ito, Proc. 10th Int. Conf. Chemical Vapor Deposition, Vol.**87-8**, 645 (1987).
[127] I. Hirase, T. Sumiya, M. Schack, S. Ukishima, D. Rufin, M. Shishikura, M. Matsuura and A. Ito, Tungsten Workshop III, 133 (1988).
[128] J.R. Creighton, J. Electrochem. Soc, **136**, 271 (1989).
[129] J.R. Creighton, J. Vac. Sci. Technol., **7**, 621 (1989).
[130] R. Foster, L. Lane and S. Tseng, Tungsten Workshop III, 159 (1988).
[131] E.K. Broadbent and W.T. Stacy, Solid State Technol., dec, 51 (1985).
[132] C.M. McConica and K. Cooper, J. Electrochem. Soc., **135** (1003 (1988).
[133] R. Chow, S. Kang, W.R. Harshbarger and M. Susoeff, Tungsten Workshop II, 137 (1987).
[134] D.R. Bradbury and T.I. Kamins, J. Electrochem. Soc., **133**, 1214 (1989).
[135] R.H. Wilson and A.G. Williams, Appl. Phys. Lett., **50**, 965 (1987).

Problems with aluminum.

[136] J.W. Diggle, T.C. Downie and C.W. Golding, Chem. Rev., **69**, 365 (1969).
[137] C.W. Borgmann, C.P. Larrabee, W.O. Binder, H.L. Burghoff and E.H. Dix Jr., Corrosion of Metals, 1 th Ed., Am. Soc. Metals, Cleveland (1946).
[138] K. Hinode, I. Asano and Y. Homma, IEEE Trans. Electr. Dev., **36**(6), 1050 (1989).
[139] D.S. Gardner and P.A. Flinn, IEEE Trans. Electron Dev., **35**(12), 2160 (1988).
[140] N. Hirashita, M. Kinoshita and T. Ajioka, J. Electrochem. Soc.,

135(12), 3159 (1988).

Tungsten as interconnect.

[141] M. Brassington, M. El-Diwany, R. Razouk, M. Thomas and P. Tuntasood, IEEE Trans. Electron Dev., **36**(4), 712 (1989).
[142] C. Kaanta, W. Cote, J. Cronin, K. Holland, P. Lee and T. Wright, IEDM, 209 (1987).
[143] R. A. Chapman, R. A. Haken, D. A. Bell, C.C. Wei, R. H. Havemann, T.E. Tang, T. C. Holloway and R. J. Gale, IEDM, 362 (1987).
[144] T. Bonifield, S. Crank, R. Gale, J. Graham, C. Huffman, B. Jucha, G. Smith, M. Yao, S. Aoyama, Y. Imamura, K. Hamamoto, H. Kawasaki, T. Kaeriyama, Y. Miyai, N. Nishimura and M. Utsugi, Symp. on VLSI Technol., pag. 101 (1988), San Diego Ca.
[145] P.I. Lee, J. Cronin and C. Kaanta, J. Electrochem. Soc., **136**, 2108 (1989)
[146] Y. Nakasaki, K. Suguro, S. Shima and M. Kashiwagi, J. Appl. Phys. **64**, 3263 (1988)
[147] C. Arena, S. Deleonibus, G. Guegan, P. Laporte, F. Martin and J.L. Pelloie, Proc. 17th European Sol. St. Device Res. Conf., ESSDERC'87, ed. G. Soncini and P. Calzolari, p. 41 (1988), Elsevier, Amsterdam.

Stress in CVD-W.
[148] R. Blumenthal, G.C. Smith, H.Y. Liu and H.L. Tsai, Tungsten Workshop IV, 65 (1989).
[149] S. Sivaram, S. Chen, D. Liao, R. Shukla and D. Fraser, Tungsten Workshop III, 407 (1988).
[150] D. S. Campbell in "Handbook of Thin Film Technology", L.I. Maissel and R. Glang ed., Chapter 12, McGraw-Hill, New York (1983).

Deposition of copper.

[151] Y. Arita, Tungsten Workshop V, 335 (1990).
[152] J.A. Kelber, R.S. Blewer, R.D. Lujan ans G. Gutierrez, ibid, 345 (1990).

[153] Y. Hazuki, H. Yano, K. Horioka, N. Hayasaka and H. Okano, ibid, 351 (1990).

[154] P.L. Pai, C.H. Ting, C. Chiang, C.S. Wei and D.B. Fraser, 359 (1990).

[155] C.K. Hu, M.B. Small, F. Kaufman and D.J. Pearson, ibid, 369 (1990).

Selective tungsten encapsulation of aluminum.

[156] H. Yamamoto, S. Fujii, T. Kakiuchi, K. Yano and T. Fujita, IEDM, 205 (1987).

CVD equipment.

[157] P. Burggraaf, Semiconductor Internat., June, 65 (1989).

[158] C. VanLeeuwen, Semiconductor Internat.,Jan., 68 (1990).

[159] M. E. Bader, R. P. Hall and G. Strasser, Solid State Technol., 5, 149 (1990).

Thermal Diffusion.

[160] R. B. Bird, W.E. Stewart, E. N. Lightfoot, "Transport Phenomena", pag. 574, John Wiley publ., 1960.

[161] L. Gillespie, J. Chem. Phys., 7, 530 (1939).

[162] I. Goldhirsch and D. Ronis, Physical Review A, 27, 1616 (1983).

[163] C.R. Kleijn, in "Tungsten and Other Refractory Metals for VLSI Applications V, pag. , 1990, Mat. Res. Soc. Pittsburg Pa.

[164] T.L. Ibbs, Physica IV, 10, 1133 (1937).

[165] G. Wahl, Proceed. Euro CVD V, J.O. Carlson ed., pag. 88 (1985), Uppsala Sweden.

[166] R.C. Reid, J.M. Prausnitz and T.K. Sherwood, "The properties of Gases and Liquids", McGraw-Hill Book Company, NY (1977).

References

Tungsten sources.

WF_6:
[167] R. Hogle and K. Aitcheson, Tungsten Workshop I, 225 (1985).

WCl_6:
[168] C.M. Melliar-Smith, A.C. Adams, R.H. Kaiser and R.A. Kushner, J. Electrochem. Soc., **121**, 298 (1974).
[169] N. Hashimoto and Y. Koga, J. Electrochem. Soc., **114**, 1189 (1967).
[169a] A. Hårsta and J.O. Carlson, Workshop on Tungsten IV, 245 (1989).
[169b] A.M. Shroff, High Temp.-High Pressures, **6**, 415 (1974).

$W(CO)_6$:
[169c] M. Diem, M. Fisk and J. Goldman, Thin Solid Films, **107**, 39 (1983).
[170] L. Kaplan and F. d'Heurle, J. Electrochem. Soc., **117**, 693 (1970).
[171] G.J. Vogt, J. Vac. Sci. Technol., **20(4)**, 1336 (1982).

Wafer temperature.

[172] D.S. Blair and G.L. Fowler, J. Vac. Sci. Technol. **A 6(6)**, 3164 (1988).
[173] D. R. Wheeler, W.R. Jones Jr and S.V. Pepper, J. Vac. Technol., **A 6 (6)**, 3166 (1988).
[174] J.E.J. Schmitz, J.L.G. Suijker and M.J. Buiting, Tungsten Workshop IV, 211 (1989).

Tungsten characterisation.

[175] M.J. Verkerk and I.J.M.M. Raaijmakers, Applied Optics, **25**(20), 3602 (1986).
[176] E.D. Palik (ed.), "Handbook of Optical Constants of Solids", Ac. Press, Inc., Orlando, Florida, (1985).
[177] T.I. Kamins, D.R. Bradbury, T.R. Cass, S.S. Laderman and G.A. Reid, J. Electrochem. Soc., **133**, 2555 (1986).
[178] L.I. Maissel in Handbook of Thin Film Technology, chapter 13, L.I. Maissel and R. Glang ed., McGraw-Hill Book Co., New York (1970).
[179] L. Eckertova, "Physics of Thin Films", Plenum Press, New York (1986).

[180] A.J. Learn and D.W. Foster, J. Appl. Phys., **58**, 2001 (1985).

[181] W.A. Metz, J.E. Mahan, V. Malhotra and T.L. Martin, Appl. Phys. Lett., **44**, 1139 (1984).

[182] W.A. Metz and E.A. Beam, Proc. Multilevel VLSI Interconnect Conf., IEEE, 357 (1985).

[183] C.A. van der Jeugd, A.H. Verbruggen, G.J. Leusink, G.C.A.M. Janssen and S. Radelaar, Tungsten Workshop V, 267 (1990).

Alternative plug processes.

[184] M.T. Welch and C. Garcia, Proc. V-MIC Conf., 450 (1986).

[185] J.L. Yeh, G.W. Hills and W.T. Cochran, Proc. V-MIC Conf., 95 (1988).

[186] N. Kobayashi, M. Suzuki and M. Saitou, IEEE Trans. E.D., **37**, 577 (1990).

[187] J.E.J. Schmitz, R.C. Ellwanger, O. Scherbaum and W.J.M. Havermans, Tungsten Workshop IV, 129 (1989).

PECVD of Tungsten.

[188] J.K. Chu, C.C. Tang and D.W. Hess, Apll. Phys. Lett., **41**, 75 (1982).

[189]C.C. Tang, J.K. Chu and D.W. Hess, Solid State Techn., march, 125 (1983).

[190] S. Tsuzuku, E. Nishitani, M. Nakatani and A. Shintani, Proceed. Electroc. Soc., **87-4**, 24 (1987).

Conversion of poly crystalline silicon in tungsten.

[191] N. Kobayashi, M. Suzuki and M. Saitou, IEEE Trans. E.D., **37**, 577 (1990).

[192] J.G. Black, D.J. Ehrlich, J.H.C. Sedlacek, A.D. Feinerman and H.H. Busta, IEEE E.D. Lett., **7**,422 (1986).

Photo enhanced CVD of tungsten.

[193] T.F. Deutsch and D.D. Rathman, Appl. Phys. Lett., **45**, 623 (1984).
[194] Y.S. Liu, C.P. Yakymyshyn, H.R. Phillip, H.S. Cole and L.M. Levinson, J. Vac. Sci. Technol. B, **3**, 1441 (1985).
[195] S. Tsuzuku, E. Nishitani, M. Nakatani and A. Shintani, Proceed. Electroc. Soc., **87-4**, 24 (1987).

Tungsten gates.

[196] N. Yamamoto, S. Iwata, N. Kobayashi and T. Terada, Proc. of 15th Conf. Solid State Devices and Materials, 217 (1983).
[197] N. Kobayashi, S. Iwata, N. Yamamoto and T. Terada, Proc. Symp. on VLSI Technol., 92 (1983).
[198] S. Iwata, N. Yamamoto, N. Kobayashi, T. Terada and T. Mizutani, IEEE Trans. Electron Devices, **ED-31**, 1174 (1984).
[199] N. Yamamoto, H. Kume, S. Iwata, Y. Yagi and N. Kobayashi, J. Electrochem Soc., **133**, 401 (1986).
[200] N. Kobayashi, S. Iwata, N. Yamamoto and N. Hara, Tungsten Workshop, 159 (1987).
[201] M. Wong and K. C. Saraswat, Symp. on VLSI Technol., 1989, Kyoto, p101, The IEEE Electron Devices Society.
[202] L. Krusin-Elbaum, M.O. Aboelfotoh, T. Lin and K.Y. Ahn, Thin Solid Films, **153**, 348 (1987).
[203] Y. Pauleau, Solid State Technol., febr., 61 (1987).

Selective tungsten on implanted oxide.

[204] W.A. Hennessy, M. Ghezzo, R.H. Wilson and H. Bakhru, J. Electrochem. Soc., **135**, 1730 (1988).
[205] D.C. Thomas, N.W. Cheung, I.G. Brown and S.S. Wong, Tungsten workshop V, 233 (1990).

References

Burried tungsten.

[206] S.S. Tsao, R.S. Blewer and J.Y. Tsao, Appl. Phys. Lett., **49**(7), 403 (1986).

Contamination.

[207] M.A. George and R.J. Haney, Workshop VI, 63 (1991).
[208] D.A. Bell, Z. Lu, J. L. Falconer and C.M. McConica, Workshop VI, 31 (1991).
[209] R.A. Hogle, P.C. Brown, Workshop VI, 47 (1991).
[210] K.A. Aitchison, E.K. Broadbent and R.A. Hogle, Workshop III, 171 (1988).
[211] I. Hirase, T. Sumiya, M. Schack, S. Ukishima, D. Rufin, M. Shishikura, M. Matsuura and A. Ito, Workshop III, 133 (1988).

CVD-WSi$_x$.

General.
[212] S. Sachdev and R. Castellano, Semiconductor Int., may, 306 (1985).
[213] K.Y. Ahn and S. Basavaiah, Thin Solid Films, **118**, 163 (1984).
[214] B.L. Crowder, Tungsten Workshop III, 3 (1988).
[215] S. Mihara, Y. Murao and M. Kikuchi, Proc. V-MIC conf., 396 (1986).
[216] M. Azizan, R. Baptist, T.A. Nguyen Tan and J.Y. Veuillen, Appl. Surface Sci., **38**, 117 (1989).

SiH$_4$ based CVD-WSi$_x$.
[217] D.L. Brors, J.A. Fair, K.A. Monnig, K.C. Saraswat, Solid State Technology, April, 183 (1983).
[218] W.I. Lehrer and J.M. Pierce, Proc. 4th Int. Symp. Semicond., ed. H.R. Hoff, R.J. Kriegler, Y. Taheishi, The Electrochem. Soc., p. 588 (1981).
[219] K. Akitmoto and K. Watanabe, Appl. Phys. Lett., **39**(5), 445 (1981).
[220] D. Dobkin, L. Bartholomew, G. McDaniel and J. DeDontney, **137**(5), 1623 (1990).
[221] C. Bernard, R. Madar and Y. Pauleau, Solid State Technol., February,

79 (1989).

[222] S.L. Zhang, R. Buchta, Y.F. Wang, E. Niemi, J.T. Wang and C.S. Petersson, Proc. 10th Int. Conf. CVD, ed. G.W. Cullen, Vol. 87-8, 135 (1987).

[223] E.J. Rode and W.R. Harshbarger, J. Vac. Sci. Technol., **B8**(1), 91 (1990).

[224] D.L. Brors, J.A. Fair, K. Monnig, K.C. Saraswat, Semiconductor Int., May, 82 (1984).

[225] K.C. Sarawat, D.L. Brors, J.A. Fair, K.A. Monnig and R. Beyers, IEEE Trans. Electr. Dev., Vol. ED-30(11), 1497 (1983).

[226] T.E. Clark, J. Vac. Sci. Technol., **6**(6), 1678 (1988).

[227] M.D. Deal, D. Pramanik, A.N. Saxena and K.C. Saraswat, Proc. V-MIC, IEEE, 324 (1985).

[228] P.S. Trammel, Proc. V-MIC conf., IEEE, 319 (1985); W. Metz, VLSI Systems Design, Februari (1986).

[229] Y. Shioya, T. Itoh, I. Kobayashi and M. Maeda, J. Electrochem. Soc., **133**(7), 1475 (1986).

[230] M. Kottke, F. Pintchovski, T.R. White and P.J. Tobin, J. Appl. Phys. **60**(8), 2835 (1986).

[231] F.K. LeGoues, F.M. d'Heurle, R. Joshi and I. Suni, Mat. Res. Soc. Symp. Proc., Vol. 54, 51 (1986).

[232] F.M. d'Heurle, F.K. LeGoues, R. Joshi and I. Suni, Appl. Phys. Lett., **48**(5), 332 (1986).

[233] Y. Shioya and M. Maeda, J. Appl. Phys., **60**(1), 327 (1986).

[234] T.P. Chow and A.J. Steckl, IEEE Trans. Electr. Dev., **ED-30**(11), 1480 (1983); G.C. Chern, J. Peterson and C. Ha, Proc. V-MIC Conf., june, 301 (1985).

SiH_2Cl_2-WSi_x.
[235] J.B. Price, S. Wu, Y. Chow and J. Mendonca, Tech. Proc. Semicon West, 1986.

[236] T.H. Tom Wu, R.S. Rosler, B.C. Lamartine, R.B. Gregory and H.G. Tompkins, J. Vac. Sci, Technol., **B6**(6), 1707 (1988).

[237] D. Srinivas, G. Raupp and J. Hillman, Tungsten Workshop V, 407 (1990).

[238] S. Selbrede, Semicond. Int., August, 88 (1988).

[239] A.S. Inamdar and C.M. McConica, Tungsten Workshop V, 93 (1990).

References

[240] T. Hara, T. Miyamoto and T. Yokoyama, J. Electrochem. Soc., **136**(4), 1177 (1989).

Fluorine content.

[241] M. Fukumoto and T. Ohzone, Appl. Phys. Lett., **50**(14), 894 (1987).

[242] Y. Shioya, S. Kawamura, I. Kobayashi, M. Maeda and K. Yanagida, J. Appl. Phys. **61**(11), 5103 (1987).

[243] P.J. Wright and K.C. Saraswat, IEEE Trans. Electr. Dev., **36**(5), 879 (1989).

[244] R.C. Ellwanger, K.D. Prall, D.R. Malinaric, R.W. Williams, J.E.J. Schmitz and E.I. Bromley, Tungsten Workshop VI, 335 (1991).

Stress.

[245] M. Shishino, T. Nishiwaki, A. Mitsui, S. Imanishi and M. Shiraishi, Proc. V-MIC Conf., IEEE, 447 (1989).

[246] Y. Shioya, T. Itoh, S. Inoue and M. Maeda, J. Appl. Phys., **58**(11), 4194 (1985).

[247] Y. Shioya, K. Ikegami, M. Maeda and K. Yanagida, J. Appl. Phys., **61**(2), 561 (1987); T. Hara et al., Appl. Phys. Lett., april issue (1991).

Step coverage.

[248] J. Hillman, J.B. Price, B. Triggs and M. Aruga, Tungsten Workshop VI, 329 (1991).

[249] G. Raupp, T.S. Cale, M.K. Jain, B. Rogers and D. Srinivas, Surface Coatings, February, (1990).

Pretreatment.

[250] R.S. Nowicki and P. Geraghty, Tungsten Workshop VI, 345, (1991); P.A.M. van der Heide, M.J. Baan Hofman and H.J. Ronde, J. Vac. Sci. Technol., **A7**(3), 1719 (1989); B.E. Deal, M.A. McNeilly, D.B. Kao and J.M. deLarios, Proc. Fall Meeting Electrochem. Soc., Hollywood, Florida, Oct. (1989); A. Izumi, K. Touei, A. Yamano, Y. Chong and N. Watanabe, Proc. CVD-XI, Electrochem Soc., K.E. Spear and G.W. Cullen ed., p. 425 (1990).

References

<u>General.</u>

[251] J.I. Ulacia, S. Howell, H. Körner and Ch. Werner, Appl. Surf. Sci., **38**, 370 (1989).

[252] D.S. Campbell in "Handbook of Thin Film Technology", McGraw-Hill Inc., ed. L.I. Maissel and R. Glang, page 12-3 (1983).

[253] R. Glang and L.V. Gregor, ibid, chapter 7, page 7-52 1983).

[254] L.P. Valdes, Proc. IRE, **42**,420 (1954).

[255] E.K. Broadbent, A.E. Morgan, R. Ellwanger, J.M. Flanner, B. Coulman, D.K. Sadana and L. Gutai, Tungsten Workshop IV, 259 (1989); A. Deneuville, M. Benyahya, M. Brunel, J.C. Oberlin, J. Torres, N. Bourhila, J. Paleau and B. Canut, Appl. Surface Sci., **38**, 139 (1989).

[256] M. Hansen, in "Constitution of Binary Alloys", McGraw-Hill, New York, 2nd edition, (1958).

[257] R. de Boer, R. Boom, W.C.M. Mattens, A.R. Miedema and A.K. Niessen, in "Cohesion in Metals (Transition Metals Alloys)", North-Holland Physics Publ. (Elsevier Sci. Publ. B.V.), Amsterdam, (1988).

[258] W. Kern and C.A. Deckert, in "Thin Film Processes", J.L. Vossen and W. Kern ed., Academic Press, New York, p. 401 (1978).

[259] H. Körner, K. Koller, U. Seidel, B. Willems and S. Kampermann, Tungsten Workshop VI, 369 (1991).

[260] P. M. Smith and M.O. Thompson, Tungsten Workshop V, 311 (1990).

[261] S. Dushman, "Scientific Foundations of Vacuum Technique", J.M. Lafferty ed., Wiley and Sons, N.Y., 1962.

[262] R.S. Muller and T.I. Kamins, "Device Electronics for Intergrated Circuits", Wiley and Sons, N.Y., 1977.

[263] E.K. Broadbent, J. Vac. Sci. Technol., **B5**(6), 1661 (1987).

[264] A. Sakamoto, H. Tamura, M. Yoshimaru and M. Ino, Proc. Appl. Phys. Soc., Japan, Fall Meeting, paper 28A-SZD-22, p. 670 (1990).

[265] S. Sivaram, P.J. Ficalora and K.C. Cadien, J. Appl. Phys., **58**(3), 1314 (1985); V. Malhotra, T.L. Martin, M. T. Huang and J.E. Mahan, J. Vac. Sci. Technol. **A2**(2), 271 (1984).

[266] J.E.J. Schmitz, unpublished results.

[267] R. Chow, pers. comm.

[268] F. Cotton and G. Wilkinson, in "Advanced Inorganic Chemistry", J. Wiley and Sons, New York, p. 957 (1972)

[269] Creare Inc., Hanover, NH.

[270] R.A. Bowling and G. B. Larrabee, J. Electrochem. Soc, **136**(2), 497

(1989).

[271] P. Huggett, Tungsten Workshop I, 233 (1985);

SUBJECT INDEX

V

via definition, 4

void, 2, 5, 21

W

WAl_4, WAl_5, WAl_{12}, 119

WAs_2, W_2As_3, W_4As_5, 120

WB_4, W_2B_5, WB, W_2B, 120

weight gain method, 117, 193

wet etching

       silicide, 191

       tungsten, 118

worm hole, 60

WCl_6, 56, 111, 112, 180

$W(CO)_6$, 56, 111

WF_6, 56, 61, 111, 113

WF_4, 74, 83, 113

WO_3, 58, 114

WOF_4, 58, 114, 120

WP_2, WP, W_3P, 120

WSi_x, 14, 171

WSi_2, 120, 180

W_5Si_3, 66, 120

X

X-ray, 66, 164, 190,

APPENDIX. UNIT CELLS OF W AND WSi$_2$

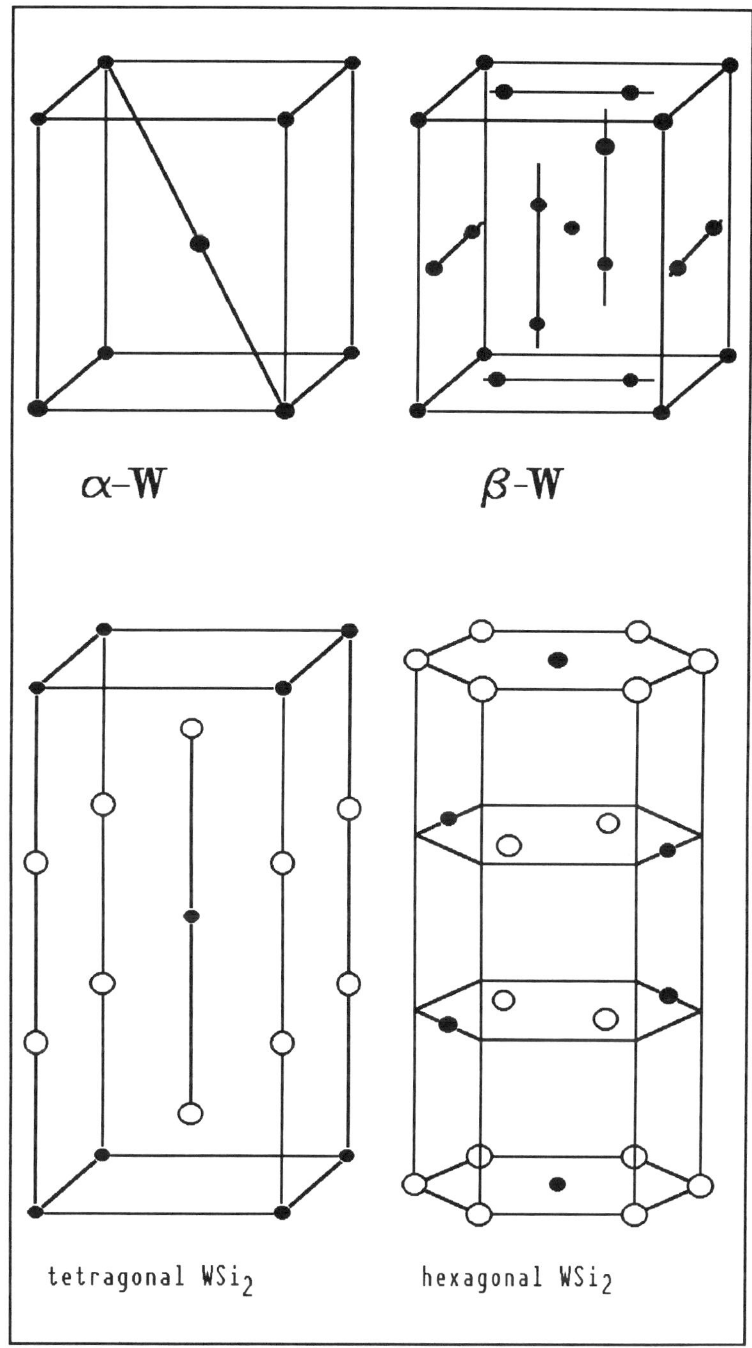

Unit cells of W and WSi$_2$. See for more details Morcom et al. [116] and Chow et al. [234]. ● = W, ○ = Si.